In Search of Reality and the Nature of Consciousness

J. Oliver Linton

Version 3.0

ISBN 978-1-291-38697-4

Contents

Part I

Part II

Exploring Further

First Walk

It was a perfect day. We had just climbed up the steep northern flank of Catbells and thrown ourselves down on the grass for a rest. Before us lay a vista of unparalleled beauty. Derwent Water lay at our feet, flecked with the sails of boats and sparkling in the sunshine. In the hazy distance the great giants of Skiddaw and Blencathra stood proud over the pretty little town of Keswick

I was fishing about in my rucksack for the flask of coffee; my companion was chewing a piece of grass and gazing up at the sky when he suddenly said:

"Have you ever thought how incredibly improbable this is?"

"What? A sunny day in the Lake District?" I quipped.

"No. The fact that *we* are *here, now*. Look," he said, turning towards me, propped up on one elbow, "here we are, you and me, right at this minute, sitting on a speck of cosmic dust, spinning through a vast empty void at a million miles an hour... I mean, it beggars belief doesn't it?"

I had to agree with him but I let it pass as I poured out the coffee.

"And another thing" he said, "Why me? Who am I anyway? If I didn't exist, would any of this exist either?" sweeping his hand across the fabulous view. "What's the point of it all?"

"Here, have some coffee."

"Who was it who said 'All the world's queer save thee and me – and even thou'rt a bit queer.'?" he asked a little while later.

"I don't know." I replied.

"Well, whoever it was, he had a point."

"I am not sure how to take that."

"Well, what I mean is that you look like me (Not as handsome, of course); you talk like me; you seem to behave like me; but you aren't me. How can I be sure there is a 'me' inside 'you'?"

"You can't. But I am in the same position, you know."

"So you say." he replied, doubtfully.

"Look, you have got to start somewhere. You can't live your life with all these doubts. You have got to draw up a list of those things which you are just going to have to accept and take it from there."

"That's all very well," he said, "but all these modern theories of Relativity and the like seem to have kicked all my common sense ideas about the space and time into touch. And Quantum Theory – well, not even the scientists themselves can agree on what it all means!

"I mean, in Newton's day, a spade was a spade; space and time were separate; the angles of a triangle added up to 180°; everything had a cause etc. Things were simple back then."

"Not so simple." I said. "Many people at the time could not stand the idea that space and time went on for ever. Nor could they stomach the idea that massive objects could influence one another across vast distances of space instantaneously, without there being any intermediate connection between them. And, of course, there were huge scientific problems to solve like the precise nature of matter, not to mention electricity, magnetism and light."

"So, do you think that science has solved all these problems? Can science give a convincing answer to the question of why we are here? Can it explain all this beauty? Can it explain how we are conscious? And, above all, is there any room in your scientific world view for *free will*?"

"You don't ask much, do you?" I said, "but I will say this: science may have no answer to the question of *why* we are here, but I truly believe that we are very close to knowing *what* is here and *how* it came to be that way. What is more, I believe that Quantum Theory does open up the possibility of knowing what consciousness is and how it can exercise what we call *free will*."

"Well, I wish you would explain it to me."

"Are you sure?"

"We have got all day ahead of us," he said, getting up from the ground. "Just keep it as simple as you can, there's a good chap."

A List of Assumptions

There can be few people on the planet who have not, at some time, suddenly been overwhelmed by the really big questions –

- Who am I?
- Who are you?
- What exists?
- What is space?
- What is time?
- What is consciousness?
- Is the future predestined or do I have a choice in the matter? Do I have free will?
- Why does anything exist at all?

At such times we are overawed by the immensity of the questions we are posing and for many, the experience admits of only one answer – GOD. Surely, the argument goes, only an omnipotent being who is both creator and mentor can provide meaningful answers to questions such as these.

I can understand this reaction, but the human race has come a long way since this hypothesis was first suggested and, by rational thought and careful experiment, the methods of science have presented us with a set alternative answers which might be summarised in the following way:

'I' am a collection of atoms and molecules arranged in a fabulously complicated way as a result of billions of years of evolution;

'You' are the same;

Out there is a universe of particles, waves and fields which obeys strict mathematical laws; which came into being a finite time ago and which has evolved over time into the state we see it today.

Science currently does not have satisfactory answers to the mysteries of consciousness and free will.

There is no answer to the question "Why?". The universe just "Is".

It must be readily admitted that the last sentence is, to many, rather unsatisfactory, not to say defeatist, but if that is the way it is, that is the way it is.

It must also be admitted that, as yet, science has little or nothing to say about consciousness but it seems to me that Quantum Theory has opened up a door which might lead to an understanding in the future. There is still much we can learn about the workings of the human brain and it may be possible in the future to investigate the phenomenon of consciousness scientifically, to be able to determine definitively which creatures are conscious and which are not and, perhaps, even to create artificial consciousness.

In addition, whereas in the past the existence of free will has seemed to be incompatible with the determinism of classical physics, it is now at least possible to imagine ways in which the powerful sense that we all have of being able to choose between different courses of action can be made consistent with the laws of physics without the action chosen being either completely determined in advance or merely random.

These are the ideas which I wish to explore in this little book.

But first, I wish to delve into some related and deep issues which, like the problem of consciousness, lie on the borders of science and philosophy. To some extent, the resolution of these issues will be largely determined by a certain set of metaphysical assumptions which must be made and it will be as well to state these assumptions at the outset in the form of a kind of creed which we might call Objective Realism. Any reader who violently objects to any of these assumptions may feel free to toss this modest little book into the nearest waste-paper basket and save himself any further wasted time but I hope that many readers will find that the list chimes in with a fundamentally common-sense view of reality.

1. The universe is a realm called space-time which has three spatial dimensions and one temporal dimension.

2. Within time and space there exist entities which we variously describe as atoms, fields, strings etc. etc.

3. An EVENT is a partial description of some of the entities described above in a particular region of space-time with a definite temporal coordinate.

4. There exists a point on the temporal axis which is unique and which is called THE PRESENT. Only events whose temporal coordinate is THE PRESENT exist; events in THE PAST have existed; events in THE FUTURE will exist . In other words, events in The PAST cannot be changed; events in THE FUTURE have not yet been determined.

5. Some events with different temporal coordinates are invariably related by mathematical law. Such events are said to be CORRELATED. Under certain circumstances (depending on the prevailing physical theory) an event A may be said to CAUSE a correlated event B (whose temporal coordinate must be larger).

6. Consciousness is a pattern of events occurring in a brain.

7. Conscious entities can exhibit FREE WILL – that is to say, there are some events which are neither caused by any previous event, nor are they random.

To be strictly accurate, the statements listed above are not so much assumptions as goals. That is to say, they are things which I *want to* believe rather than things which I *actually* believe. As we shall see, the discovery of Relativity and Quantum Theory in the first quarter of the 20th century has cast some doubt on the validity of some of these assumptions – numbers 4 and 5 in particular – and it is one of my main objects to explore the extent to which these assumptions can be upheld in the face of these discoveries.

So, having stated my assumptions, the BIG questions that stare us in the face are these:

1. **What is SPACE and what is TIME?**
2. **What constitutes REALITY?**
3. **What is CONSCIOUSNESS and do conscious beings possess FREE WILL?**

In what follows, I shall try to present my arguments at several different levels. Firstly, in a fanciful discussion between myself and an imaginary walking companion; secondly as a summary of my views; and thirdly, in more detailed discussions of the relevant ideas, experiments or theories mentioned in the summary. Hopefully, in this way, readers can follow the argument without too much distraction at first and can follow up unfamiliar aspects of the discussion at their leisure by following up the cross-references given in the text.

First Walk continued...

An hour later we were munching our sandwiches on the summit of High Spy admiring the spectacular view southwards towards Great Gable and the Scafell Range. As I was scanning the scene through my binoculars, I saw a tiny grey dot heading towards us at speed.

"There's a Tornado coming over Sty Head!" I exclaimed.

"Don't be silly. It's a cloudless sunny day." said Alan.

"No, look, there it is. It's a plane."

And soon we could both see the soundless aircraft banking steeply to make the turn into Borrowdale. Then, as it shot past us on our left hand side, we heard the sound, apparently travelling at the same speed but a mile behind the real plane.

"Isn't it funny how the sound travels behind the plane." commented my companion. "It is almost as if there were two planes. One real and one audible, one travelling behind the other."

"Actually, there are *three* planes." I said. "One real, one *visual* and one audible. Because of the finite speed of light, the visual plane travels behind the real one just like the audible one."

"How far behind?" Alan asked.

Light travels a million times faster than sound. If the audible aircraft is a mile behind the real one, the visual aircraft is 1 millionth of a mile behind.

"2 millimetres." I said after doing a quick mental calculation.

"Hey, that's quite a lot! I had no idea it would be so much. I thought light travelled so fast you could ignore it completely."

I must admit, I was a bit surprised too.

"I wonder what the world would look like if light travelled at the speed of sound." Alan mused. "I suppose it wouldn't make much difference to objects nearby but if I looked at Big Ben, say, through a telescope, it would appear to tell the wrong time."

"Yes, and all sorts of other things would look odd too. If you were watching a long train coming towards you it would appear to be stretched and as it passed by it would shrink!"[1]

"Isn't that what Relativity is all about?"

"Well, yes and no. The effects we have just mentioned are simply due to the finite speed of light. The slowing of clocks and the shrinking of rulers predicted by Special Relativity are additional effects due to the strange fact that observers in relative motion always measure the same speed of light. But let's not get into that just now. Let's just accept that two different observers have a different perspective on things, but they are still fundamentally observing the same world."

"But doesn't Relativity sometimes change the *order* in which events occur?"

"Yes, that's true," I replied, "but that is no more puzzling than saying that, to us, High Scawdel over there is to the left of Dale Head – but if we were sitting on the top of Great Gable, Dale Head would be to the left of High Scawdel."

"Yes, I can see that it works in *space* but surely you can't change the order of things in *time*. If High Scawdel *caused* Dale Head then it can't be that, to someone else, Dale Head *causes* High Scawdel."

"That's a really good point, but I assure you Relativity is perfectly consistent with causality provided that *no influence can pass from one object to another faster than the speed of light.*"

---

1   See the footnote on page 126 for an explanation of this effect

"But didn't I read somewhere the other day that, if you make an observation on a proton or something over here, it can instantly affect the state of a proton on the other side of the universe."

"Where did you read that? In *The Sun* perhaps?"

"Don't patronise me. It was in *Physics World* actually."

"OK. OK." I said. "You are quite right. And it is a puzzle which worries a lot of clever people too. Fortunately, the effect cannot be used actually to send a signal faster than light so it is not really inconsistent with Relativity. It is a worry all the same. Good Heavens! Look at the time. We had better be on our way if we want to get back before the pubs shut!"

"Wouldn't it be nice," said Alan, "If there was a big knob somewhere which you could turn to alter the speed at which Time goes. I mean – I wish this day could go on for ever, it is so glorious up here. Then tomorrow, back at work, I could speed up Time so that it passed more quickly."

"What a wonderful crazy idea! Perhaps you could even rotate the knob in the opposite direction to make Time go backwards."

"Now it is you who is being ridiculous."

"Yes. Sad to say, the whole idea is ridiculous. But it raises some interesting issues all the same."

"What issues?"

"Come on, we must be off."

1: What is SPACE and what is TIME?

Fundamentally, all our knowledge about the universe in which we live comes to us through our senses. More specifically, through our *conscious* senses. And perhaps the most compelling sense we have when we are *conscious* is the sense of the existence of time. It is almost possible to close your eyes and imagine yourself somewhere else, outside the universe even; but it is quite impossible to imagine yourself some*when* else – because the very process of imagining something compels you to imagine it *now*.

To put it another way, if consciousness exists (we can be sure of that if nothing else) then every conscious being defines a unique point on the temporal axis. Naturally, this argument does not require that every conscious being has to share the same point on the temporal axis as every other being but Special Relativity puts severe constraints on the range of points which two conscious beings can regard as simultaneous. (See: *The four dimensions of Space-time* on page 104)

In addition to the existence of time, our senses also compel us to take regard of the *passage* of time. And we can do this too without reference to any external stimuli because our minds allows us to imagine X *now* and Y sometime *later*. What is more, our memory permits us to put X and Y in a specific order. Just bring to mind a pink dragon. Got that? Now imagine the thing that you are going to imagine next. You can't do that because you don't yet know what you are going to imagine next. Now imagine an honest car salesman (difficult, I know). Now which did you imagine *first*? The dragon or the salesman?

When we open our eyes and perceive a three (spatially) dimensioned world around us in which events appear to happen in definite order, we are (in general, rightly) convinced that the inanimate objects whose universe we share also exist in the same temporal dimension as ourselves and (at least for objects close to us and moving at relatively slow speeds) share the same PRESENT.

Once we have decided that (within the limits imposed by Special Relativity) the PRESENT is a universal feature of the world around us, the temporal axis is immediately divided into two – the PAST and the FUTURE. Once again, the theory of Special Relativity, far from blurring the distinction between the two, makes the difference absolutely explicit. (see: *Past and Future in Special Relativity* on page 105)

Two questions now arise. How fast does the PRESENT move along the temporal axis and why does it move from PAST to FUTURE and not in the other direction?

Many people, faced with the difficulty posed by this last question, have rejected this idea of the PRESENT completely and consider the PAST and the FUTURE to have equal status – in effect considering the whole of space-time as existing, as it were, simultaneously. A few have gone down the road of inventing a second temporal dimension to regulate the speed of the first – then a third to regulate the second, then a fourth – and so on into an infinite regress.

I am not satisfied with either of these positions. My conscious mind tells me emphatically that a unique point in time called THE PRESENT actually exists and that the events on one side of this point have existed in THE PAST and other events do not yet exist and lie in THE FUTURE. This is the common-sense view and it will take a lot more than a cleverly worded conundrum to cause me to doubt it. My answer to the question – how fast does time move? – is simple. It doesn't matter. I don't care if a second lasts a million years in some other temporal dimension; to me, a second lasts, well, precisely one second.

You could ask the same question about length. How long is a metre? How do you know that the whole universe and everything in it didn't just double in size a minute ago? The answer is, of course, that it doesn't make the slightest difference. Even if the universe did double in size a minute ago, my metre wide table is still the same length as the metre ruler lying on it. In the same way, even if Time speeded up, seconds would still remain seconds.

No, the problem with time is not the rate at which it flows but that it flows at all; and in one direction only – though again, why we should have much of a problem with this is, to me, more of a mystery than the problem itself. Nevertheless, the question is still worth pursuing and many, many books have been written about the arrow of time both scholarly and fictional.

It seems to me that there are three different questions which we could pose at this point:

1.   What would the (classical) world look like if time were suddenly to reverse?

2.   Is there any evidence of any temporal asymmetry in the laws which govern fundamental particles?

3.   Can we deduce the direction of time from the fact that causes always precede effects?

Lets take them one by one.

Reversing Time

Just suppose you were lucky enough to posses a big red switch carrying the labels 'FORWARD' and 'REVERSE' which had the remarkable effect of instantaneously and exactly reversing the velocities of every atom, photon and fundamental particle in the universe. What would happen if you were to throw the switch? Obviously the Earth would reverse in its orbit and the teapot which your mum had at this instant accidentally knocked off the table would stop falling and would, instead, rise gracefully and reposition itself back where it belonged. Also, all the photons which were at that moment streaming away from the Sun would suddenly find themselves travelling towards it and any electron which had recently been emitted from a radioactive atom would find itself on a course which would lead to its recapture by that atom. But would radioactivity cease to happen? Would the Sun cease to shine? Would human beings get younger every year? Would the Solar system wind itself back to the time when it was a cloud of dust around a proto-star? Would the universe collapse back into a Big Crunch?

Our starting point is the remarkable fact that nearly all the laws of Physics are completely symmetric with respect to time. The list includes Newton's laws of motion and his law of gravity; Maxwell's laws of electromagnetism and, by implication, all the laws which govern the behaviour of light. The list also includes Einstein's laws of Relativity and even Schrödinger's wave equation which is at the heart of Quantum Theory.

If you take a film of a collision between two snooker balls, or of a comet weaving its way through the solar system, or of light wave bouncing off some mirrors, or even of an electron in a quantum state and run it backwards through your projector, the resulting scene will not disobey the laws of physics. Indeed you may not even be able to tell that the film is being run backwards.

Even if you take a film of the 'break' in a game of snooker, the ripples produced by a stone thrown into a pond, a bursting balloon or a radioactive atom decaying, the reversed film will look very unlikely *but it will still not disobey any of the laws of physics* –

– Except the Second Law of Thermodynamics.

So what is this wonderful law which is asymmetric with respect to time and which will give Time its Arrow? In one of its many guises it can be expressed as follows: 'In any isolated system of sufficient size, there is a quantity called entropy which always increases.' So there you have it. All you have to do is set up such a system, watch it for a while and if entropy increases, you know time is running forwards and if entropy decreases, it is running backwards.

Sorry – it is not quite as simple as that.

It is my contention that if you were suddenly to throw the time switch into reverse, *entropy would still increase*. It follows from this that you cannot, in fact, use the second law to 'prove that Time runs forwards' or to 'determine which way Time is running' because whichever way Time goes, you always see the same thing – namely, entropy increasing. Why is this?

Well, the reason is that the Second Law of Thermodynamics is not actually a physical law at all. You could say that it is a tautology. Once you have defined entropy, it is bound to increase whichever way time is running.

To illustrate what I mean, consider your mothers cherished teapot which, you may remember, she knocked off the breakfast table. Suppose that this time you are not quite so quick at flicking the time switch and the teapot has already hit the floor and the handle has come off. All goes well at first. At the instant you flick the switch, all the velocities of all the atoms and molecules in the universe reverse their directions and you watch as the teapot and the handle fly back towards each other. You confidently expect the two pieces to rejoin and for the reunited teapot to rise gracefully back onto the table again - but this is not what happens at all. Quantum theory tells us that there is an inherent uncertainty in the positions and momenta of all microscopic particles and this uncertainty is sufficient to ensure that the atoms of the teapot and the handle cannot fuse back together exactly as they were before. In fact, even the minutest differences in the positions of the atoms when they hit each other will prevent the necessary chemical bonds from forming and the handle will just bounce off the body of the pot. The handle-less teapot now has a different centre of gravity than it had when it was falling and in consequence, hits the edge of the table as it rises causing it to fall to the floor again, shattering it into a yet more pieces. Measurements of the entropy of the system would show that, after a very brief fall, entropy would continue to rise just as before!

To put it another way, in any universe which contains the tiniest bit of randomness in its fundamental laws, The Second Law of Thermodynamics is symmetric with respect to time – just like Newton's Laws of Motion.

If this is the case, how can we explain the obvious asymmetry of all the everyday events which occur around us? The answer is quite simple. We live in a universe which has a finite past and which began in a highly ordered state (i.e. a state of very low entropy). Why it began like this is not a question which, I believe, science can currently address, let alone answer; but given that this is how it began, the universe has subsequently been evolving into a more and more disordered (and likely) state ever since. When you see a film showing teapots being miraculously mended or snooker balls converging and arranging themselves into a neat triangle, you know that the film is running backwards, not because the situation disobeys any of the laws of physics but because you are watching a system which is spontaneously moving from a disordered state to one which is more ordered (and therefore much less likely). (For a more detailed discussion of these issues see: *Randomness, Order, Disorder and Entropy* on page 110 and *The Second Law of Thermodynamics and the Arrow of Time* on page 114.)

But the second law of thermodynamics does not apply to absolutely all physical systems which seem to have a preferred direction in time. If you take a film of the ripples produced when you throw a stone into a pond, the second law definitely applies to the stone – but not to the ripples. When you throw the time switch, the circular ripples will suddenly reverse their direction and if you were shown just this part of the film you would probably be inclined to say the the film was running backwards even though the passage of a wave through a frictionless medium does not involve a change in entropy.

Why is this? Well, circular ripples converging on a point are very unusual. If you came across such a thing happening in real life you would have to suppose that an army of children all round the pond were deliberately disturbing the water in synchronism. Another alternative would be that the pond was bounded by a circular reflector and that the ripples had originally been caused by a stone thrown into the centre. Outgoing ripples, on the other hand are very common because ripples contain energy whose origin is (usually) concentrated in a small place. Consider the difference in the ripples produced on the surface of a pond by a heavy rainstorm and the ripples left behind in a swimming pool out of which a class of energetic children has just emerged. If you were shown a short reversed clip of these events you would tell immediately that the former was reversed but the random nature of the second situation would look the same either way.

What we can infer from this is that, although entropy itself remains unchanged when a wave spreads out from a source, the reason why a reversed wave is recognisable as such is because we can (usually) infer that it had its origin in a low entropy situation in the past (like the stone falling into the pond). If we consider systems whose entropy is already maximised (like the swimming pool) we cannot tell the difference between forward and reversed time.

Fundamental Particles

What about experiments on fundamental particles? Do they show any unexpected asymmetries?

The answer to that is yes. Instead of asking ourselves, does time run forwards or backwards, let us ask a related question: is the universe left or right-handed? Richard Feynman distilled this question is a particularly graphic way. He asked us to imagine that, at some time in the future, we made contact with intelligent beings on a nearby star. Through a lengthy process of education, we learn to communicate with each other and to send pictures of our planets to each other. But there is a problem. How do we know which way to scan the pictures we receive so that left is left and right is right? How could we tell them the difference between clockwise and anti-clockwise? How would our new friends know which hand to hold out when the time eventually comes to shake hands?

Amazingly, in 1957 it was discovered that when a spinning nucleus decays by beta decay, when you look in the direction of travel of the emitted electron, the nucleus is (usually) spinning anti-clockwise. By repeating this experiment, our alien friends could determine exactly what we mean by words like left and right, clockwise and anti-clockwise.

This experiment exposed what is known as the Violation of Parity. It turned out, though, that that was not the end of the story because it was realised that anti-nuclei would spin in the opposite direction! So our alien friends would only get the right message if they were made of matter not anti-matter! In other words, the experiment would look the same if you reversed both Parity (P) and Charge (C). This is known as CP symmetry.

Since then, other experiments have been carried out that suggest that CP symmetry is also violated. This means that we could find out if the aliens were made of matter or anti-matter – but only if time went forward! (i.e. CP symmetry can be preserved if we reverse Time as well. This is known as CPT symmetry)

It is even conceivable (though it has not happened yet) that a particle reaction could be discovered which is genuinely asymmetric in time as well, thus violating even CPT symmetry. This would enable us to tell if alien time went forwards or backwards relative to ours. But this doesn't mean that alien time *could* go backwards. All it means is that we would be able to predict the outcome of our alien friends' experiments. (It would, however, be impossible to tell them the results of our predictions because I do not see how we could possibly communicate with them if their time went backwards.) More to the point, we still could not deduce from the experiment whether time *in our own universe* was running forwards or backwards.

Cause and Effect

We can trace the origins of the temporal asymmetries in the two first examples we considered (the second law of thermodynamics and outgoing radiation) back to the fact that we live in a rather special universe which began in a fantastically low entropy state and which has a long way to go before it reaches thermodynamic equilibrium. If we lived in a universe which was in the latter state then all films of events occurring in the universe would look identical whether run forward or backwards. Time would cease to have any meaning. In fact, there wouldn't actually be any events to record let alone film cameras or film crews to record them.

So if we cannot look to either physical phenomena or particle physics to explain why time runs forwards instead of backwards, is there some *logical* or *philosophical* reason? Could we, for example, point to the fact that, logically, causes always *precede* their effects and that if we were to encounter an effect which preceded its cause, we would know that time was running backwards[2].

For example, the reason why the golf ball flew down the fairway was because it was struck by a golf club. The impact of the club on the ball is the cause and the subsequent motion of the ball is the effect. Cause precedes effect therefore the club must have hit the ball before the ball flew off and time must have flowed from the former event to the latter.

The argument above is so full of holes, hidden assumptions and circular reasoning that it is barely worth the effort of pulling it apart. All I will point out is that if you reversed the event exactly the same argument could be used to maintain that the momentum of the incoming ball was the reason why the club bounced backwards.

In fact, at the microscopic level, there is no such thing as cause and effect. Things just *happen*. When we say things like 'the earthquake caused the building to fall down' or 'the bright star which Johannes Kepler saw in 1604 was caused by a massive supernova explosion which occurred 20,000 years before' we are summarising a vast chain of happenings stretching from one event to the other. All the intervening happenings are causeless and effectless because, at the microscopic level, *all events, by definition, happen at the same time* (an event is something which happens at a certain time). When atom A collides with atom B, atom B collides with atom A simultaneously. If you hold that causes must always precede effects, it makes no sense to say that the change in atom B is caused by the collision with atom A. They simply

---

2 You may know that Special Relativity permits the order of certain events to be reversed but fortunately the circumstances under which this can happen are rather special and do not affect the issue of causality. See: ***Special Relativity and Causality*** on page 109

collide.

When we talk about an earthquake as the cause of a falling building, what we are really saying is this: *as a result of long experience with similar situations and also theoretical considerations, I am of the opinion that if the earthquake had not occurred, the building would not have fallen.*

Two things need to be said about this. Firstly, the statement concerns theoretical concepts ('earthquakes', 'buildings', 'supernovae' etc.) which have no part to play in any physical law but which do play an important role at what we might call a higher level of law. Like 'temperature' and 'pressure' these concepts can be used to predict and explain events which it would be impossible to explain using the fundamental laws of physics alone. They are what are called 'emergent properties' (see page 190 for more details of **Emergent and Transcendent Properties**) and all statements about cause and effect necessarily involve emergent properties. It would be extremely unwise to pin your hopes of explaining the Arrow of Time on the basis of such ill-defined concepts.

Secondly, the statement has the form of what is known as a 'counterfactual conditional'. Although very useful, such statement have very dubious logical validity and we should not rely on such statements to prove that falling buildings could not cause an earlier earthquake to happen or converging photons to cause a supernova.

So if there is nothing in either Physics or Philosophy which absolutely prevents either of these unlikely events happening, all we can do is ask:

Does Backward Time make any sense?

All this discussion about problems with the Arrow of Time is, I believe, a massive red herring. It is like asking whether a football could roll backwards. It is in the nature of a football to roll and the direction in which it rolls is forwards. It simply makes no sense to ask if it could roll backwards. If you reverse its motion, you find that it is still rolling, well, forwards. If you had two footballs, it would make sense to ask if they were rolling in the same direction or not but as we have seen, although the existence of other universes in which time flows backwards is not ruled out by the laws of physics, we could not communicate with its

19

inhabitants. The idea of time flowing backwards either makes no sense or it makes no difference. Either way, we should stop worrying about it!

Past, Present and Future

Once we have decided that Time exists and the direction it moves in is *defined* as being from the past towards the future, all we have to discuss is the status of these latter concepts. The common sense approach is to consider that events in the past HAVE existed and events in the future have YET TO exist but it does not seem to make sense to say that the past events exist NOW. Indeed, if it did, it would make just as much sense to say that future events also exist NOW. This train of thought has lead many to suppose that the past and the future had the same existential status. What then?

There are two possibilities. If both the past and the future already exist, then the world is like a video which we can run forwards or backwards as we please. In such a world there is no room for free will as there is no possibility of the future turning out to be anything different than it is. What is more, there is no room for causality either because we can with equal validity maintain that the past is caused by the future as the other way around. While I may be prepared to modify my views on free will and causality, I am not prepared to abandon them completely.

The other possibility is that neither the past not the future exist, only the present. On this view, the world is like a video with only one frame in it. May be the frame is constantly changing but we would be totally unaware of this because all our conscious thoughts, memories included, would belong exclusively to the present.

Again, I think these issues are largely misconceived. We use the word 'exist' to describe an important feature of the world we live in. Some things, like Dolphins, exist; others, like Dragons, do not. Now I ask you – do Dinosaurs exist? If you simply say: 'No, Dinosaurs do not exist' you are missing something important about them. I believe that Dinosaurs existed in the past and that this fact is true NOW. I also believe that the truth or falsity of the statement that Dragons will exist in the future is at present *undecided*. It is not a question of whether the future and/or the past exist (now); what is important is whether *statements about the future or past are true, false or undecided* (now).

The Limits of Space and Time

There is one other question which must now be raised in connection with the nature of space and time – can either have a beginning or an end? The question is particularly pertinent in the case of time because modern theories of cosmology postulate that time did indeed have a beginning; what is more, we can even claim that time began 13,600,000,000 years ago – which raises the obvious (but, as we shall see, misguided) question "What happened before the beginning of time?"

Now it is not quite enough simply to retort "Don't be silly! There was no time before the beginning so you can't use the word 'before'". There are much better reasons why the question cannot be asked. And to see why, let us ask the same question about space – a concept which seems much less paradoxical than time. Let us suppose that we are told by a respected scientist that space is not infinite and that it has a definite edge. We ask him facetiously what does the edge look like? Does space end in brick wall? If so, what is on the other side? He smiles wryly and says – come with me and I will show you. Rather reluctantly, we accompany him into his spaceship and after a long journey he invites us to put on our space suits and take a walk outside. To our surprise, looking back the way we have come we see a familiar universe full of stars and galaxies, but in the opposite direction – nothing. Just blackness. The professor tells us that we are floating at the edge of space. In fact, we are just one metre away from the very edge itself. "Ridiculous!" I say, "There may be no more galaxies beyond this point but I can see that there is plenty more space!" and I start to reach out my arm to prove my point. "Wait!" cries the professor, "That may be dangerous!" and I quickly withdraw my arm. "Watch this." the professor says, drawing a long metal rod out of his life support pack with which he proceeds to probe the edge of space. As the tip of the probe gets closer to the edge, the rod is stretched out until eventually it breaks and bits fly off the tip, disappearing into the black void with a shower of sparks. It is clear from his demonstration that there is no possibility of any of us exploring the region on the other side of the edge because we would get ripped to shreds before we crossed the boundary. "Where do the bits go?" I ask the professor and he tells me that they are effectively lost from our universe for ever, only the mass and charge remain measurable. I can hardly believe him; the 'edge' looks perfectly normal, just completely black. No, not totally black. Now that I look more closely there seems to be a reddish glow marking the boundary. "What's that faint reddish glow that I can see?" I ask. "Oh, that's light emitted by objects as they crossed the edge millions of years ago. It has taken that long for the light to climb back into our universe and has been greatly red-shifted in the process. Come on, it's time to get back on board."

As we return to the spaceship, the professor tells us that the edge of space is not all like that. There are other regions where the edge is quite different. And so, a little while later we are once again floating at the very edge of space. As before, the 'edge' looks like an empty void but I can't help noticing that the edge itself appears to be littered with bits of what looks like paper stuck all over the place. Once again, the professor gets out his probe. This time, instead of the probe getting stretched longer and longer, it gets compressed shorter and shorter. The more the professor extends the rod, the more compressed it becomes until it is clear that the tip is never going to cross the invisible boundary. Once again, it dawns on us that, it is simply not possible to investigate the space 'beyond' the edge – if it exists at all. It simply is not part of the universe we live in.

"What are all those bits of paper?" I ask the professor who tells me that they are objects that are falling out of our universe. As they get closer and closer to the edge, they travel more and more slowly and so, to us, they seem to us to be frozen in time. I notice that one of them looks like a pocket watch. Yes it *is* a pocket watch! As I look at it more closely I see that its hands are spinning round and round wildly. I point this out to the professor who replies "That's because, as you get closer to the edge, time goes faster and faster. From the point of view of the pocket watch, it looks as if the universe is infinite but from our perspective, we can see that the universe has an edge over there." (For a further discussion of these ideas see: ***Black Holes and Brick Walls*** on page 124)

With our brains reeling from all this nonsense, we return to the ship where the professor assures us, much to our relief, that none of our current theories about the shape of Space require that it has an 'edge'. The universe we live in is thought to be either infinite or 'finite but unbounded'. (See: ***General Relativity and the shape of the Cosmos*** on page 128) On the other hand, the Big Bang Theory does suggest that Time, while having an infinite future, may have a finite past. But just as with my two 'edge of space' scenarios, there are good reasons why it is impossible, even in principle to explore or even know anything about the time 'before' the Big Bang. The reason is simply that, at the time t=0, all the properties of matter and energy which we can, in principle, measure such as temperature and density become infinite. All our current theories of physics, Relativity and Quantum Theory included, assume that when a system evolves from a state A to a state Z, it does so because of a causal chain which extends all the way through B, C, D etc. But if, at some point in the chain, everything blows up to infinity, the causal chain is broken and it becomes quite impossible to *explain* how the system gets from A to Z. In the same way, it is quite impossible to explain how the universe evolved from a state before the Big Bang to a state after it; the best thing we can do is describe its evolution from a point arbitrarily close to the origin. The question 'what caused the Big Bang ?' is, in my opinion, not only unanswerable, it is meaningless.

Incidentally, if it were to come about that, at some point in the future, all the measurable quantities in the universe became zero, exactly the same argument would apply. The universe would (obviously) cease to exist and the question 'what happens then?' would be equally meaningless.

Conclusions

Although it has often been said that the theories of Special and General Relativity have revolutionised our ideas of Space and Time, it turns out that none of our listed assumptions have to be changed radically. We still have basically three dimensions of Space and one of Time; there is still a well defined past and a well defined future which all observers can agree on (even though observers in relative motion may disagree a bit about what exactly constitutes the present); most important of all, the concept of *causality* comes through unscathed.

The only new idea that we have to add is the possibility that Space and Time might be finite. We can reject speculations about what happens beyond the edge of the universe or before the beginning of time as fascinating but ultimately meaningless fantasies.

Second Walk

I was frankly terrified.

We had set set out from the Old Dungeon Ghyll car park in fine weather and now, two and a half hours later, we had gained the wide shoulder between Great End and Broad Crag where we had our coffee. The path seemed obvious and I set off at a good pace knowing that our ultimate objective, the summit of Scafell Pike, was perhaps only half an hour away. Soon, however, the going became rougher. What had once been a wide, well-trodden path degenerated into a mad scramble over a chaotic jumble of rocks with little to distinguish the route. What is more, a cloud, which a moment before had been but a light wisp on the mountainside suddenly enveloped us in its clammy grip.

"Are you OK?" I shouted to my companion, Alan, who was making heavy weather of the stones a dozen yards behind.

"I'm fine." I heard him reply as I plunged ahead into the clag.

"You still there?" I said a moment later.

Silence. Only the mocking sigh of the wind among the stones.

"Where has the bugger got to?" I muttered to myself. "Oh – there he is." I thought, as I caught a brief glimpse of a figure striding off through the mist over to one side. So I turned back to the task in hand and started off again. I noted what seemed to be a cairn in front of me and headed for that – but when I get there, I was hard pushed to tell if it was a cairn or just a random pile of stones. Next a huge buttress of rock loomed up in front of me which I didn't recognise. I was beginning to get a bit panicky at this stage. "Do I go left or right?" I asked myself. I really had no idea. I went right.

A moment later and I knew I was truly lost. The summit of Broad Crag in a mist is one of the most terrifying places I know. It consists of a pile of angular blocks, none smaller than a suitcase, which are awkward to clamber over and excruciatingly painful to slip off. There is absolutely nothing about the rocks to tell you if anyone else has ever stepped on them before and nothing to tell you what direction you are heading. On all sides there are fearsome crags falling away into the depths of Wasdale and only a narrow col connecting it to the shoulder of the Scafell Pike and safety.

As I say. I was terrified.

But just at that moment, salvation arrived. A sudden, brief tear in the cloud revealed the summit of the Pike itself with its familiar walled shelter and attendant bevy of walkers. Instantly I knew where I was and which way I had to go. "Thank God!" I thought. And then a moment later: "Christ! What about Alan? I wonder what has happened to him?"

15 minutes later I was climbing up the last few steps to the summit cairn with dread in the pit of my stomach only to come face to face with Alan himself, calmly eating a sandwich.

"What ever happened to you?" he said.

"Me? What ever happened to you? I thought you were following me!" I exclaimed.

"I was. Only, it turned out to be someone else."

"You could have fallen off a cliff!"

" Oh no. He seemed to know where he was going."

"That's not the point. He could have fallen off a cliff too. Come to think of it – I could have fallen off a cliff. Did you think of that?"

"Not really," he said, "Anyway, you're here now and that's all that matters. Here – have a sandwich."

I sat down beside him, vastly relieved to be safe again, but not entirely convinced that being on the summit was 'all that mattered'.

As we consumed our lunch together the fog blew away and soon we could see where we had come across the shoulder of Broad Crag on our way up.

"I wonder where I got lost." I mused, trying to trace the path I might have taken across the boulders.

"Does it matter?" asked Alan, now tucking into some cake.

"I would just like to know, that's all." I said. "I must have gone *somewhere*."

"Perhaps not." said my companion. "When you are in a fog like that, you are in a different world. Time seems to stand still somehow. The fog presses on your eardrums and nothing seems *real* any more. The rocks and stones begin to look as if they are stage props – you know, like they were made of polystyrene or something. Perhaps you wandered about all over the place and only became real again when the sun came out. Perhaps *you* really did fall off a cliff and are now clinging to a ledge with a broken leg hoping that your mobile phone hasn't been smashed; meanwhile, *I* have chosen to go down a different road to reality and am contentedly eating this biscuit, talking to a different you. Who knows?"

"What utter crap you say sometimes." I commented.

But then, after a moments silence, a thought suddenly struck me.

"On second thoughts, I think you have just said something really important – profound even."

"Oh yeah? Well, there is a first time for everything – or so they say."

" Look: there is a lot of debate at the moment – well it has actually been going on for nearly a century now – about what Quantum Theory has to say about what is real and what is not real. There are so many bizarre experiments which seem to rule out the possibility that there is just one reality out there and scientists are generally divided between two opposing camps. There are those who follow the teachings of one of the great founders of Quantum Theory, Niels Bohr, who said that we can never know what actually happens down at the atomic level; the only thing that we can know is what our instruments tell us. This is the 'you're here now and that's all that matters' approach, otherwise known as the Copenhagen Interpretation of Quantum Theory."

"On the other hand" I continued, "there are those who like to think that, at the atomic level, everything possible happens at once, and there are indeed worlds where I fall off a cliff and worlds where the moon is made of green cheese."

Alan snorted. "You can't be serious!"

"I agree with you," I replied, "but the idea has gained popularity since Hugh Everett proposed it in the 1950's and it is now considered to be at least as orthodox as the Copenhagen Interpretation."

"Well, I don't think much of either idea."

"Nor do I. But something you said has given me another idea. What if reality is *suspended* for a while. What if, while I was lost in the fog, reality was put on hold while several possibilities were tried out. In one world I fell off a cliff; in another I climbed up to the top of Broad Crag. When the sun came out, it was 50:50 which possibility would become real but once the decision was made, not only did reality reassert itself, so did the history which led up to that reality become real too."

"Is this a new idea?"

"Well, not entirely. There are plenty of people who are looking for a mechanism whereby a quantum system 'collapses' or 'decoheres' but I don't think anyone has used the term 'suspended reality' before and to me the idea makes more sense than either denying the existence of reality completely or postulating an infinite number of simultaneous realities."

"Yes, I can see that. You can tell me a bit more if you like – but let's get off this wretched mountain first!"

2: What is REALITY?

Reality has come in for a hard knocking in the last century and opinions range from 'Reality is an illusion – it doesn't exist' to 'There are an infinite number of different Realities which coexist and can even interfere with each other.' The problem started when Albert Einstein showed that different observers in relative motion had different but equally valid perspectives on reality – but the rot really set in when Niels Bohr articulated his 'Copenhagen Interpretation' of Quantum Mechanics in the 1920's. Since then, thinkers who, like me, want to believe in a single objective reality have repeatedly come up with apparently impossible consequences of the theories of Relativity and Quantum Theory – such as the possibility that space travel could make your father younger than you (see: *The Twins Paradox* on page 133) or that cats could be both dead and alive at the same time (see: *Schrödinger's Cat* on page 145) – only to have experimental proof of these apparent impossibilities thrown back in their faces.

It is now impossible to continue to hold the notion that there is only one single objective reality which everyone shares. But neither is it necessary to abandon the concept altogether. Two observers in relative motion may disagree about a lot of things – but there is a lot about the universe which they observe which they *can* agree on. Likewise, we may not be able to say which slit the electron went through – but we can predict with great accuracy where on the fluorescent screen the electron can and cannot land (see: *The Double-slit Experiment* on page 146).

The problems posed by Relativity pertain mainly to the nature of Time and can all be resolved by a simple change in coordinates. Two observers in relative motion or near a black hole can be said to have a different perspective on reality which may cause them to disagree about certain facts like the length of their rulers or the speed of their clocks, but there is still a reality which they can still share. They will always measure the same *interval* (See: *The 4 Dimensions of Space-time* on page 104) between two events and they will never disagree on which events caused which etc. etc. These problems are insignificant to those posed by Quantum Theory.

QT poses four problems; first there is the discovery that small objects like atoms and photons have both wave-like and particle-like properties at the same time – the so-called **wave/particle duality** (See:

Waves and Particles on page 149); next there is the existence of **superposed states** e.g. the possibility that a cat might be both dead and alive at the same time (See: *Schrödinger's Cat* on page 145); then there is the **measurement problem** – the problem of how superposed states transform themselves into a single reality (See: *The Measurement Problem* on page 155); and then there is the **EPR paradox** – the idea that an observation made on one entity here can instantaneously influence the properties of a different entity over there (See: *Entanglement* on page 157).

Many, many ideas have been suggested in order to try to make sense of the experimental results but nearly all of them reject the idea of a single objective reality. The only suggestion which comes close to retaining this axiom is the idea that every particle comes with a family of 'runners' who dart about at infinite speed, gathering information about the current state of the rest of the universe, reporting back and telling the particle what to do. For example, while an electron is minding its own business passing through one slit, its runners (call them virtual particles or pilot waves if you like) will discover the other slit and tell the electron not to land on the forbidden areas of the fluorescent screen (See: *Pilot waves and Virtual Particles* on page 163). Or again, a polarized electron whose spin is 'up' will be told by its runners that its positron colleague on the other side of the solar system has just had its horizontal polarization measured and will, in consequence, flip over sideways (See: *Bell's Theorem and the Aspect Experiment* on page 165).

No; this doesn't seem likely, and this idea has not enjoyed much popularity among scientists either, (though it is currently making a bit of a comeback) and yet – what is the alternative?

Dozens of experiments have conclusively shown that the behaviour of sub-atomic particles is inconsistent with the idea that they have well-defined properties which exist even when they are not being measured. To get a flavour of the kinds of things that quantum physicists have to explain, have a look at *The Twin Monkey Paradox* on page 167. Pragmatic Quantum Theorists who simply want to get results push the problem under the carpet by claiming that there is no objective reality at all other that that embodied in their equations. This is the standard Copenhagen Interpretation of QT (See: *The Copenhagen Interpretation* on page 161). At the other extreme, there are those who claim that all possible realities exists simultaneously (but that each conscious mind only goes down one of the possible routes). This is the Many-Worlds Interpretation (See: *The Many-Worlds Interpretation* on page 162). I don't like either of them. I passionately want to believe that there is one objective reality that we all share – but I am prepared to accept that, on a sub-atomic scale, reality may not have quite the same character as it is perceived to have on an everyday scale.

The issue comes into sharp focus when we consider the theories that have been put forward over the centuries to explain the behaviour of light. In particular, I should like to consider the commonplace phenomenon of specular reflection – the reflection of light off an ordinary mirror. Every schoolchild is familiar with the Law of Reflection – *angle of incidence = angle of reflection* – and every schoolchild is told that this behaviour comes about because light bounces off a mirror like a football off a wall. Essentially this is Newton's corpuscular theory of light (See: *Newton's Corpuscular Theory of Light* on page 134).

A few years later, the student will be told that reality isn't like that at all. Light is, in fact, a continuous electromagnetic wave – and he will be shown a proof that waves obey the same law of reflection as well (See: *Huygens' Wave Theory of Light* on page 135).

A few years later still, the student, now an undergraduate, will be told that the football analogy was not far wrong after all and that light is a stream of particles called photons which have momentum and this explains why there is a minute reaction force on the mirror when the photon reflects off it (See: *The Photon Theory of Light* on page 136).

Another professor, in a different lecture theatre, will, however, be explaining to his third-year students that these photons are actually wave packets, about 30 cm long and a few thousand wavelengths wide and this is the reason why photons do not always travel in precise straight lines (See: *The Wave-packet Theory of Light* on page 139).

Then, in a post-graduate course on Quantum Theory, the idea will be suggested that when the photons leave the source they just disappear and are replaced by a 'probability wave' which bounces off the mirror and which governs the place where the photon reappears – if you bother to look, that is (See: *The Copenhagen Interpretation* on page 161).

Elsewhere in the building, another respected scientist is arguing that these 'probability waves' are a lot of nonsense; what actually happens is that the photon, on leaving the source, splits into in infinite number of 'potential photons' which, like infants released into a playground at break time, rush out into space exploring every possible behaviour – even those that seem to violate fundamental laws of physics like travelling round in circles or going backwards in time. These photons carry a clock which pulses at a fixed frequency and when they meet another potential photon whose clock is oscillating out of phase with theirs, both photons annihilate each other. The lecturer asks his students to imagine that all the infants in the playground are wearing flashing LEDS in their trainers; when two infants whose LED's are out of phase collide into each other they disappear in a puff of smoke. Naturally this will happen very frequently and the playground rapidly empties – but not everywhere. Over in the far corner of the playground, at precisely the point where a line drawn from here to the playground wall and back to the schoolroom door obeys the law of reflection, there is a large knot of infants whose LED's are all flashing in synch. Triumphantly, the lecturer bangs his bongos and announces that this is why light obeys the laws of reflection! (See: *Feynman's theory of Light* on page 141)

During the course of as many years, students are presented with 6 different explanations of the reflection of light! No wonder students of Physics get so confused!

So which one is correct? Are they all correct? Well, yes – in a sense – they are. Are none of them correct? Well, yes – that too. Which one best describes what light really is? Well, all of them, really – it depends what aspect of light you want to describe.

One way out of this dilemma (or, more accurately hexlemma!) is to claim that none of the explanations here proposed are descriptions of what light really is, but mathematical models which serve simply to predict the behaviour of light in certain circumstances without making any attempt to explain it. This is the position adopted by adherents of the Copenhagen interpretation of QT. I find this position unacceptable. I want my science to provide explanations as well as numerical predictions. I am prepared to accept that our theories may be incomplete and that my understanding of the theories even more deficient – but where a theory fits the observed facts within its own defined domain, I want to think that the theory gives us genuine insights, if not complete answers, as to the nature of the reality it describes.

For example, Newton's theory of gravity (the idea that there exists a force between two massive objects which acts instantaneously over the space between them) has conclusively been shown to be false. But that does not mean that Newton's theory has nothing to say about the nature of gravity. Einstein's General Theory of Relativity is easily shown to boil down to Newton's law as long as the velocities of all the bodies in question are much less than the speed of light. In other words, within the low speed domain, bodies act *as if* they are acted on by an instantaneous force acting between them. In fact, I will go further than this. I really see no reason why we have to insert the italicised *as if* at all. If Einstein's theory predicts that planets will behave *as if* they have a force acting on them then why not just say that the force actually exists and that Einstein's theory explains, in detail, the origin of the force?

To take another example, does the Coriolis force (the force which, in a rotating system, causes objects to veer off course sideways) exist or not? The official view is that the Coriolis force is a fictitious force and objects in a rotating system only act *as if* they are acted on by a sideways force. If, however, you have ever tried to walk across a rapidly rotating carousel or watched a satellite video of a cyclone developing you will have been left in no doubt as to the reality of this force.

Of course, some theories are genuinely proved to be false. No-one believes that the planets are carried round in their orbits on crystal spheres any more. But where a theory such as Newton's theory of gravity is found to be an approximation to a more comprehensive theory, it is perfectly acceptable to maintain that the concepts which that theory relies on are just as real now as they were in Newton's day.

So Newton's force of gravity still exists – it is just not the whole story.

When we come to consider the competing theories about the nature of light, we come across a difficulty though. Whereas General Relativity is clearly on a higher level than Newton's theory in that you can deduce the latter from the former but not vice versa, the wave and particle theories of light (by that I really mean Huygens' and Feynman's theories) are on an equal footing and it can be shown that, mathematically, they are identical. We cannot say that light is really made of particles which behave *as if* they are waves; nor can we say the light is a wave which behaves *as if* it was a stream of particles. Somehow we have to accept that light is both a wave and a particle at the same time.

Nor is the problem confined to the nature of light. Electrons also have a dual nature and almost every interpretation of Quantum Theory predicts that, under the right circumstances, even macroscopic objects will show the weird properties that come about because of the dual nature of particles and waves: cats that are half dead and half alive, for example; footballs that veer off course when kicked through a door; encrypting devices that self destruct as soon as an intruder looks inside them etc. etc. Unpalatable though these ideas are, even objective realists like myself are being forced to accept that reality is not quite what it seems to be and that, under certain conditions, reality either ceases to exist, or that multiple realities can exist at the same time.

We seem to be between a rock and a hard place.

Which is it to be – multiple realities? or none? Many Worlds? or Quantum Limbo?

It may be as well, at this point, to distinguish between *Quantum Mechanics* and the *interpretations* of Quantum Theory. Quantum Mechanics is a mathematical formalism which enables practising scientists to use Quantum Theory to make valuable predictions is specific circumstances. There are, in fact, several different formalisms which have been devised to do this but they can be shown to be mathematically equivalent. The most important of them is probably Schrödinger's Wave Equation (See: *Schrödinger's Wave Equation* on page 151). Now it does not matter whether you subscribe to the Copenhagen Interpretation or are a Many-Worlds enthusiast. You can still use the same mathematics and will get the same result. We only have to choose between the two if we want to try to understand *why* the mathematics works.

Actually, I do not believe that we are forced to choose between these alternatives. Like the wave/particle duality itself, we can accept both as equivalent models of what is going on in the sub-atomic world. I don't really care if you explain the behaviour of an electron passing through a double slit experiment by either denying that it goes though either of the slits, or that it goes through both. As long as your mathematical formalism predicts the right results, either approach will do and we can make our decision simply on the basis of which approach makes the calculations easier, not on the basis of what we think is really happening.

What I will not accept, however, is that this ambivalent attitude to reality which we are forced to adopt at the sub-atomic level must be extended all the way up to the macroscopic level as well. There must, I believe, be a point where quantum effects cease and reality kicks in again – but where exactly is this point? When do waves turn back into particles? What causes this collapse? This is, of course, the Measurement Problem and in its solution lies the key to all of the other weird effects of the quantum world such as superposition and entanglement.

Now some of the best brains in the world are currently working on this problem and it is, perhaps, unlikely that this modest essay will contribute anything genuinely new but I would like to suggest an idea which I find satisfying and which might, in time, suggest new answers to some of the most baffling philosophical issues of our day including the nature of consciousness.

Firstly, in order to free myself from either the Many Worlds or the Quantum Limbo scenarios, let me introduce the concept of:

Suspended Reality

When an electron or a photon leaves a source and starts to propagate through space, I like to imagine that reality is 'suspended' for a while. You may prefer to think of reality as becoming meaningless or you may prefer to think of reality as splitting into an infinite number of possible realities; I don't mind. Either way we will talk about reality being 'suspended' for the duration. During this time, wavefunctions may develop, potential particles may dart about the place, God may get out his calculator and start pressing buttons feverishly, I don't care. The important thing about suspended reality is that it is temporary. Eventually reality reasserts itself and the photon or electron causes a macroscopic, potentially observable effect somewhere.

Let's look at a simple analogy. Imagine it is October 1939 at the very start of the Battle of the Atlantic. A report is flashed through to the

Navy Command Centre in London that a merchant ship has been sunk by a U-boat 50 miles West of Shetland at 0300 hours. The position and time is plotted on the operations chart in the ops room by the WAAF on duty. Time passes. At 0600, Winston enters the room. 'Where's that bloody submarine now?' he asks, gruffly. 'We don't know, sir, but, assuming he is still submerged and is travelling at less than 7 knots, it must be in this circle here' the WAAF replies, indicating a circle of radius 30 miles with her pointer. More time passes and the circle of probability gets larger, distorted now by the presence of land where the submarine obviously cannot go. The submarine has to be *somewhere*, though, so as the circle gets larger, so the probability of finding it in a particular place gradually decreases.

Then at first light a Sunderland flying boat spots a submarine on the surface recharging her batteries and reports that the submarine has been disabled with a torpedo. Suddenly, the probability wave disappears and the flag indicating the submarine's position is moved to a new, definite position of the chart.

This analogy has many beguiling features. The probability distribution certainly behaves a bit like a wave in that it spreads out with a certain maximum speed and that it can collapse in an instant. But analogies are also dangerous because they may suggest interpretations which turn out to be entirely false. In this case, we know that at all times, the submarine is actually somewhere – even if we don't know where. The position of the submarine is what is known as a 'hidden variable'. Einstein was convinced that there would be 'hidden variables' beneath the surface which would explain the strange behaviour of electrons and photons but he failed to find them. Countless subsequent experiments (See: **Entanglement** on page 157, **Bell's Theorem and the Aspect Experiment** on page 165 and **The Twin Monkey Paradox** on page 167.) have shown that the search for 'hidden variables' is doomed to failure and we have come to accept the unpalatable fact that while a photon is traversing an optical bench or an electron is orbiting the nucleus of an atom, it really is not 'there' or anywhere else either. Only the probability of its being 'there' exists.

The question now arises, what causes a system which is in a state of suspended reality to collapse into a single reality? This problem is not unique to my theory; any theory which involves the 'collapse of the wavefunction' has to address this issue as well.

One of the first suggestions (by Von Neumann in the 1930's) was that quantum collapse only occurs when the 'wave of probability' or wavefunction reaches a conscious mind. This is equivalent to saying that Moon ceases to exist (i.e. is in suspended reality) all the time you are not looking at it – an idea repudiated by Einstein (and me) as wholly unacceptable.

Another hypothesis could be that entities such as electrons or atoms undergo spontaneous collapse with a tiny but finite probability of collapse in any period of time. If we couple this with the hypothesis that, if any one one entity in a quantum system collapses, the whole system collapses, then we have a plausible theory. Small ensembles of atoms will stay in a susoension for long periods of time but the greater the number of atoms in the system the greater the chance of collapse. The main problem is that there is nothing in the theory which determines the actual 'half life' of a single atom, and it is generally agreed that any satisfactory new theory should avoid arbitrary constants at pretty well all costs.

A third, more promising possibility is that quantum collapse occurs when the mass of all the particles involved reaches a certain threshold. Individual electrons will always remain in suspended reality but as soon as one of their manifestations collides with a real atom, the atom joins the collection of particles in suspension. As time progresses, the number of atoms that are potentially involved increases and there comes a point when the threshold is reached and reality collapses back again.

At first sight this limit looks as arbitrary as the 'half life' idea proposed first but Roger Penrose thinks that the limit might emerge naturally from the, as yet unknown, theory of quantum gravity which he is working on. It may be that the quantity is related to the well known Plank mass – the fundamental unit of mass in the system of units in which the fundamental constants G, h and c are unity. This mass is of the order of a millionth of a gram which, though small, is the mass of rather a lot of atoms (See: *The Planck Units* on page 178). An objective realist like myself would prefer to confine quantum effects to just a few dozen atoms if possible!

My own preference is for something along the following lines. I concur largely with the idea that particles are real but that sometimes

36

their properties i.e. position, speed etc. are *undecided.* During these periods of suspended reality the various possible paths with the particle might take (which I call *trajectories*) are described by a mathematical function similar to Schrödinger's wave equation. (See: *Schrödinger's Wave Equation* on page 151.) As time proceeds, the wave function develops according to some, as yet unknown, rules, tracing out all the possible trajectories which the particle could take.

During this period of suspended reality, all the different possible realities or trajectories acquire more or less probability . During the process of radioactive decay, for example, the probability of finding an alpha particle outside the nucleus increases while the probability of it remaining inside decreases. A little while later, other possibilities emerge: for example the (potential) alpha particle may collide with and ionise an atom of Nitrogen. This ion might capture a passing electron or bond with a nearby hydrogen atom et. etc.. As time proceeds, a myriad of possible trajectories open up, each with an associated probability. In certain circumstances these probabilities can be calculated with great accuracy using Schrödinger's wave equation methods or Born and Heisenberg's matrix mechanics or with the aid of Richard Feynman's particle diagrams – it matters not. All we have to assume is that, however complex the range of possibilities becomes, there is, in principle, a way of calculating the probabilities of all the possible outcomes. The trouble with these current methods is that they never terminate. The possibilities go on multiplying for ever and ever.

Now, going back to our radioactive atom, let us suppose that there is a new term in the wave equation, or a new dimension to the matrix, or a new, as yet undreamed of virtual particle whose effect is to introduce an instability in the equations so that, after developing in an orderly way for a while, there comes a point where the probabilities start behaving chaotically. (See: *The Logistic Equation* on page 174) Minute differences appear in the probabilities of outcomes that an instant ago were equally probable and these differences increase at an exponential rate. Any symmetry present in the original situation is broken and the positive feedback loops present in the equations drive the amplitudes to extremes (See: *Broken Symmetry* on page 179). Eventually, one of the possible outcomes of the experiment, by chance, or rather, by the process of deterministic chaos, acquires as probability of 1. Suddenly, the probability of any other outcome automatically becomes zero and reality is restored. (See: *Chaotic Collapse* on page 177)

It is apparent that, according to this theory, what drives the collapse of suspended reality is not some arbitrary threshold, but is a direct consequence of the increasing *complexity* of the system as its potential effects spread out wider and wider into the surrounding environment. This explains immediately why two entangled photons can, in principle travel to opposite sides of the universe without losing their correlation and why a Bose-Einstein condensate of several thousand atoms can exists in a state of semi-permanent quantum superposition without collapsing into a single real state. Both these systems are effectively isolated from their environment and are mathematically very simple systems – so their quantum nature is preserved. As soon as the system is allowed to interact appreciably with the environment, the mathematical complexity of the system increases dramatically and the quantum system collapses into a single, observable reality.

Imagine a binary star system. Newtons laws of motion tell us that, providing the two stars do not collide, the system is always perfectly stable. Now toss a planet into the system. As is well known, the three body problem has no general solution. There are stable orbits, but there are also many chaotic ones too. Toss another planet into the system and the chances that one of the objects will suddenly fly off to infinity increase. The more planets you add, the more likely it is that the system will become violently chaotic.

Now it may be pointed out, quite rightly, that complexity does not necessarily imply chaotic behaviour. Our own solar system has a couple of dozen major gravitating objects in it and it appears to have been stable for billions of years – but even here, it is thought that the orbits of the planets will be impossible to predict beyond a few million years. On the other hand, chaotic behaviour can occur in quite simple systems (see: *The Logistic Equation* on page 174). But all systems which exhibit deterministic chaos are essentially *non-linear*. Now is linear and is therefore not susceptible to deterministic chaos. My suggestion is that there is something missing from our current mathematical models of quantum phenomena – that something being a non-linear term whose effects are somehow linked to the complexity of the situation being described.

I would also like to suggest that at some point in the process there appears an essential element of randomness. Schrödinger's equation is not only linear, it is continuous – that is to say it deals with variables like time and position which can take on any value. But what if time and space are discontinuous, for example on the scale of the Planck units? (See: *The Planck Units* on page 178.) Reality now becomes a series of discreet frames, like the video from a CCTV camera which only takes pictures every second. When things are happening slowly, this may not matter very much – but as the pace increases, vital information can get lost, and spurious, random effects can occur which can rapidly multiply in a chaotic system. And, of course, this fundamental randomness is exactly what we need to explain the Second Law of Thermodynamics. (See: *The Second Law of Thermodynamics and the Arrow of Time* on page 114.)

If you accept this idea – that of reality being suspended for a while followed by the collapse of the wavefunction when the system becomes sufficiently complex, all the philosophical problems with quantum theory disappear!

In the double-slit experiment, we wondered why the act of observing an electron passing through one of the slits in a double slit experiment causes the interference pattern to disappear (see: *The Double-slit Experiment* on page 146 and *The Measurement Problem* on page 155). The answer is that any signal communicated by the electron to the outside world as it passes through the slits causes a chain of potential events to occur which increases the complexity of the wavefunction to the point where it collapses. Note that it is not necessary for the electron actually to pass through the slit being monitored; it is the *potential* chain of events in the detector which causes the collapse. (See: *The Wave/Particle Duality Explained* on page 182 and *The Measurement Problem Explained* on page 184)

A number of diabolical thought experiments have been devised to explore this idea further, one of the most interesting being published by Elitzur and Vaidman in 1993. It proposes a method to test whether a photo-sensitive bomb is working or not. Obviously if you just shine a light on the sensor, either the bomb is a good one – in which case it explodes and therefore becomes useless; or it is a dud – in which case it was useless anyway. What you do is place the bomb sensor in one arm of an interferometer and a detector is a position which corresponds to a as providing a moral compass in our relations with other humans (because they, presumably have souls too) and offering the possibility of life after death (because the soul is immortal).

For those of us who do not want to believe in the religious trappings of a soul, it is still reasonable to hold the opinion that science does not yet hold all the answers in respect of how our conscious sense of self arises and that there must exist *something* which will eventually explain the riddle and of which science is either currently completely ignorant or only dimly aware (dark energy? quantum decoherence? self-organized complexity?).

It is possible to categorise the various forms of mentalism – which I here take to mean any scientific theory of the conscious mind which emphasises the importance of substances, structures or processes which are not completely described by the otherwise well understood physical processes that go on inside the neurons and synapses inside our brains – in terms of what this extra element is.

Cartesian Dualism is the doctrine that the mind is (for want of a better word) a 'spiritual' entity which overlooks the workings of the brain and which can, when required, interfere with it. (Descartes himself thought that it had a definite location in the brain and identified the pineal gland as the seat of consciousness) This 'spiritual' entity was ridiculed by the philosopher Gilbert Ryle in the 1940's who dubbed it the 'ghost in the machine' and few scientists and philosophers would own to subscribing Cartesian Dualism these days because a) there is no actual evidence for it and b) it doesn't really *explain* anything. Nevertheless, we should not throw it overboard completely. We do need to keep in mind the serious possibility that there may be something in the physical world – a substance, a structure or a process – about which we currently know nothing but which is essential to understanding the mind. Indeed, as will become clear, I myself am of this opinion and therefore could be said to be a Dualist of a sort.

Much more popular these days are theories which view the mind as an 'emergent property' of the brain in the same way that temperature is an emergent property of gas molecules, or nest building is an emergent property of a colony of ants (see *Emergent and Transcendent Properties* on page 190). This position, sometimes referred to as **Epiphenomenalism**, comes in a wide variety of guises which are minimum in the interference pattern. If the bomb is a dud, it is not capable of detecting the passing photon so the wavefunction will not collapse, interference will occur and the detector will remain silent. If, on the other hand, the bomb is a good one then reality will split into two alternative possibilities. In one, the photon hits the sensor and the bomb explodes but in the other possibility, the photon passes through the other arm of the interferometer and sometimes triggers the detector. The net result is that if you test 100 good bombs, 50% will explode and 25% tell you nothing – but in 25% of the cases you will know that the bomb is a good one without exploding it. How cool is that! (See: *The Elitzer/Vaidman Bomb-testing Experiment* on page 171)

The third mystery of Quantum Theory is the mystery of superposition – the Schrödinger's Cat paradox (see *Schrödinger's Cat* on page 145). But this mystery is a mystery no longer. Long before the cat enters the period of suspended reality, the wavefunction will have collapsed and the phial of poison will have broken – or not, as the case may be[3] (See: *Schrödinger's Cat Paradox Explained* on page 183).

Lastly, there is the question of entanglement. But it is obvious now how this issue is resolved. It really is the same as the Measurement Problem. Reality is suspended for a while until such time as it becomes necessary to decide between the several alternatives. When the collapse occurs, it is not only the case that reality starts again from the point of collapse. The bomb-testing experiment also shows that reality is *reconstructed* right back to the point where the suspension took place. When the electron lands on the fluorescent screen, it really did go through one slit and not the other. When the positron on the other side of the solar system is found to have a certain spin direction, it actually had this spin all along. The theory of suspended reality has no need for faster-than-light communication or spooky action at a distance. It is logically complete. (See: *Entanglement Explained* on page 185)

The Game of EFIL

I should like to emphasise the retrospective nature of chaotic

---

3Actually that is not strictly true. Since there is no actual threshold at which the wavefunction inevitably collapses, the theory predicts that there will always be an infinitesimally small possibility that the cat is in the alternative state so to be strictly accurate, the theory does permit a cat to be 99.999999.........9999 % alive and 0.000000.......0001% dead. I can live with that, and so (most of the time) can the cat.

collapse with an analogy.

In 1970 an article appeared in *Scientific American* describing a simple 'game' invented by John Horton Conway called the Game of LIFE. It is not really a game, it is, in fact, what is called a two dimensional cellular automaton. It is played on an infinite rectangular board whose cells can be either white (dead) or black (alive). The cells are updated simultaneously at regular intervals. The game has just two rules:

1.   A white cell turns black only if it has exactly 3 black neighbours.

2.   A black cell turns white only if it has either less than 2 or more than 3 black neighbours.

Hidden in these rules is a fascinatingly complex system whose behaviour has captivated both amateurs and professionals alike and whose properties are still not completely understood. Here is a typical sequence:

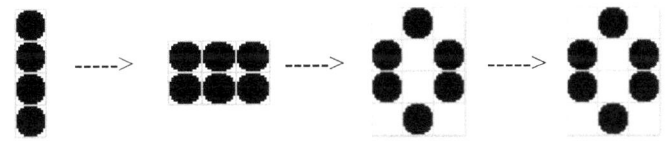

The last figure is called the 'beehive' and, as you can see, it is stable and will never change.

Here is another sequence

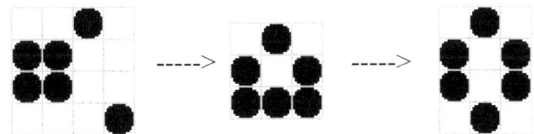

and another

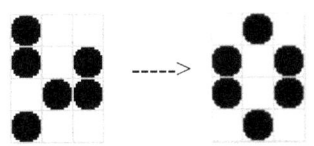

41

all of which end in the beehive.

It should be obvious that, given the rules, every pattern will generate a unique successor. It is equally obvious from the above examples that a given pattern does not have a unique predecessor. Most patterns have many possible predecessors.

Conway's game was called the Game of LIFE for obvious reasons. Patterns grow and die like living organisms and there are no known rules which will let you say whether a given pattern will live or die – you just have to try it out.

As an analogy for real life it suffers from one serious drawback. In the game of LIFE the *future* is predestined and the *past* is unknown: in real life, it is the *past* which is fixed and the *future* which is unknown.

This suggests that the game of EFIL might be a better analogy. The rules are the same, it is just played backwards. Here are some possible games of EFIL:

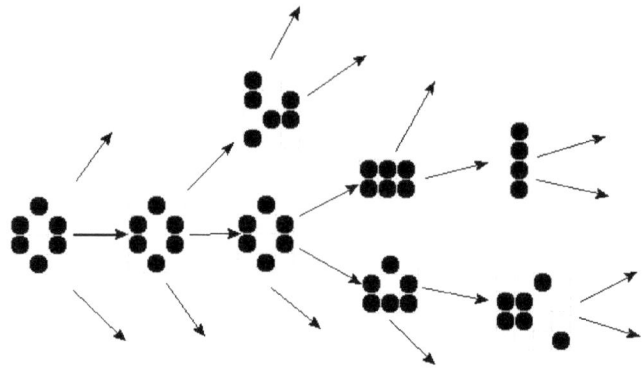

Every pattern has many possible futures but only one past.

Now let us consider the passage of an electron through a double slit experiment to the point when it hits a fluorescent screen as if it was a game of EFIL. When the electron reaches the two slits, reality splits into two possibilities S_1 and S_2 – just as the beehive can split into two possibilities.

Let us now suppose that the electron lands at one of three points on the screen X, Y or Z. (All other points being eliminated by interference.) Our two possibilities have now split further into six which we can call X_1, Y_1 and Z_1 and X_2, Y_2 and Z_2. (For example Y_2 would represent the case where the electron passed though slit 2 and ended up on the screen at position Y.) These would be represented in the game of EFIL by six different patterns. (Here we must suppose that the patterns X_1 and X_2, though different in detail, look exactly the same when viewed from our macroscopic viewpoint and the same for the pairs Y_1, Y_2 and Z_1, Z_2)

Now when wavefunction collapse occurs, one of the six possibilities become certain (i.e. real) and the others vanish. But this simultaneously determines all of its predecessors – in other words, it determines *which* slit the electron had been through, in retrospect!

It is this feature of the theory which so nicely explains the bomb-testing experiment. If the bomb doesn't explode, we know which way the photon went – even though we also know that it went both ways!

Suspended Reality – is it possible?

How does this theory of suspended reality compare with the already accepted interpretations of QT? Does it have testable consequences? Is it mathematically feasible? Is it philosophically acceptable?

First, it is necessary to recognize that the idea of 'suspended reality' is merely an different *interpretation* of the standard idea in which a quantum system can be in more than one state at once. The difference is largely semantic. When I claim that the idea can *explain* entanglement, for example, I am merely suggesting an idea which can give certain people a feeling of satisfaction. Entanglement doesn't really need explaining. If that is the way the world works, that is the way the world works.

The idea of chaotic collapse, however, is potentially a respectable scientific theory with testable consequences. For this to happen, someone will have to work out a mathematical theory which describes the chaotic collapse of the wavefunction and then devise experiments to see if the expected collapse occurs when predicted. I entertain the hope that such a theory might enable us to construct conscious machines and explain the workings of conscious brains..

It is possible that further research into quantum computers will shed light on this. So far, primitive quantum computers have been constructed which contain just a handful of qubits of information and a few quantum gates. If the theory of chaotic collapse is true, then I suspect that researchers will find the difficulties of sustaining a quantum computation in a quantum computer will increase exponentially with the number of qubits involved. If this is so, it will put a severe restriction on the potential power of such computers. The Copenhagen interpretation and the Many Worlds interpretation put no such restrictions on the complexity of a quantum computer and the difficulties should increase linearly.

There is another, still controversial, effect which may prove decisive. It is known as the Quantum Zeno effect and it claims to show that quantum collapse is a real effect by either preventing it from happening, or, in the case of the anti-Zeno effect, to stimulate its occurrence. (See: *The Quantum Zeno Effect* on page 181) It is claimed, for example, that the act of watching a radioactive atom can continually collapse the wavefunction inside the atom, thus preventing the alpha particle from ever getting out. If the effect is proven to exist, it will strengthen the case for wavefunction collapse being a real observable effect but it is not clear to me why any of the currently suggested interpretations of quantum theory should predict this behaviour. The theory of chaotic collapse does, however, require that information from the whole region surrounding the quantum system to be taken into account in determining what happens to it next, not just its current state. In this respect, the theory of suspended reality and chaotic collapse could well turn out to make different predictions from those of the orthodox interpretations and which would make its truth or falsity a matter of experiment, not conjecture.

Conclusions

In the introduction to his book *The Nature of the Physical World* Sir Arthur Eddington famously used the following words:

I have settled down to the task of writing these lectures and have drawn up my chairs to my two tables. Two tables! Yes; there are duplicates of every object about me – two tables; two chairs; two pens.

I am writing this book sitting at *three* tables! First there is the solid structure on which I rest my computer; it has weight and strength and texture. Then there is the table which is 99.999% empty space with charged particles called atoms and electrons whizzing about, interacting with each other. But then there is a third table; this one is even more insubstantial than the second; in fact, it doesn't contain any *things* at all! it is just an array of numbers in a multidimensional space which obey a mathematical law called a wave equation.

Which is the *real* table?

The answer is that they are all equally real *within their appropriate context*. Table number 1 (the solid one) is an entity which *emerges* from table number 2 (the atomic one) at a higher level rather like the way in which the face of a celebrity emerges from the, apparently random multitude of dots which make up an old newspaper photograph. Likewise, table number 2 is an *emergent property* of table number 3 (the mathematical one). (See: **Emergent and Transcendent Properties** on page 190)

Ever since Plato, philosophers have toyed with the idea that the universe is fundamentally mathematical. Galileo famously said that the book of the world was written in the language of mathematics and the discoveries of Newton, Boltzman and Maxwell seemed to prove his point. But there has always been this doubt at the back of every scientist's mind. How long can our good fortune last? Will we always be able to find mathematical structures that model the increasingly complex reality which Relativity and Quantum Theory are revealing?

In 1959 the physicist Eugene Wigner wrote[4]:

The miracle of the appropriateness of the language of mathematics for the formulation of the laws of physics is a wonderful gift which we neither understand nor deserve. We should be grateful for it and hope that it will remain valid in future research and that it will extend, for better or for worse, to our pleasure, even though perhaps also to our bafflement, to wide branches of learning.

Max Tegmark[5] has proposed a simple but radical idea which should give us renewed hope. *The universe* **is** *mathematics.* I entirely agree.

Reality has many layers but the most fundamental is mathematical.

For a few more comments on the theory I have put forward, see: *Critical Comments* on page 198.

---

4   Eugene Wigner: **The unreasonable effectiveness of mathematics in the natural sciences**. Richard Courant lecture in mathematical sciences delivered at New York University, May 11, 1959
5   Max Tegmark (February 2008). **The Mathematical Universe**. *Foundations of Physics* **38** (2): 101–150

Third walk

We were almost down when the accident happened.

That autumn morning we had set out under crisp clear skies up to Sty Head and had spent a pleasant morning 'threading' Napes Needle and scrambling about on the Great Napes buttress; we had stood on the little grassy ridge which connects the huge bastion of rock to the mountain and gazed down the length of Wastwater and we had had our lunch on the windy summit while we watched the clouds gathering in the west. Now we were picking our way carefully down Gable Beck as a snow shower whirled about us, covering the path with a slippery film of wet snow.

Suddenly, without a sound, Alan slipped and fell. All I heard was the dull thud as his head smacked against a rock. When I reached him, blood was pouring from his skull and he was showing the whites of his eyes. I quickly propped him up and clapped a (fairly) clean handkerchief to his head.

"Come on, Alan. You can't give out on me!"

I got out my mobile phone but, of course, there was no signal. I looked round for some assistance and fortunately, two other walkers were within hailing distance and had soon clambered down to join me by which time Alan was beginning to shows signs of coming round.

"He's hit his head and lost consciousness." I said. " I am sure we can get him back down to Wasdale Head so I don't think we need the

Mountain Rescue Service but he is going to need an ambulance. Could one of you go and telephone from the hotel?

"I'll go." said one of the walkers and set off down the path towards the distant valley at a cracking pace.

An hour later, we were just limping into the hotel yard when we heard the regular thump of the rotors of the Air Ambulance on its way. Soon Alan was strapped to a stretcher and headed for the Royal Infirmary in Lancaster, just 15 minutes flying time away.

By the time I had struck camp and made my way back home, night had fallen. I telephoned the hospital to make sure my friend was OK and arranged to see him the next day. There I found him, his head swathed in bandages, chirpy as a cricket. I placed my gift of a box of chocolates and a 'get well' card of Pudsey Bear on the bedside table and greeted him cheerily.

"How are you, my old chum?"

"Fine, thanks. Hey – those chocolates will come in useful."

"What do you mean?"

"Well, there is this smashing night nurse and..."

"Well I can see there is nothing wrong with you after all" I said, getting up as if to go.

"No, no, wait. Tell me what happened. I can't remember a thing."

So I sat down again and told him the story.

"Can't you really remember anything at all? Not even the ride in the helicopter?"

"No. That's a real bummer. I've always wanted to ride in a helicopter and now I have, I can't remember a thing about it."

"You wouldn't have seen anything, anyway. They gave you a sedative."

"I suppose that's true. I can't even remember getting down the mountain though I was perfectly conscious at the time."

"What happened to your arm?" I asked, noticing for the first time that his limb was strapped up as well.

"Oh, I broke my wrist as well when I fell. It's funny. I didn't notice it at the time but when I came round in the hospital it hurt like hell. They put it in plaster this morning. Gave me a general anaesthetic, thank God. I had a wonderful dream while I was under. I can't remember what was in it though."

"Girls, probably." I muttered under my breath.

"Isn't the brain a funny thing? In the last 24 hours I have been unconscious at least three times. First I was knocked silly by a blow to the head; then I fell asleep in this bed and this morning I was put under by a simple whiff of gas. What is going on inside my brain when it is conscious and how does the conscious brain differ from an unconscious one? I really would like to know."

"So would a lot of other people"

"I know that the brain contains billions of nerve cells which act like logic gates in a computer but this doesn't even begin to explain how a brain can be *conscious*."

"No, but it does explain how a brain can process stimuli coming from its senses and how it can control our limbs in response."

"Yes, but if that is all there is too it," said Alan, "we are all nothing but mindless robots, blindly following some program which has been implanted in our brains either by chance or by God or by alien beings from another planet. I don't like that idea at all."

"Neither do I. I agree with you – there must be something else going on, but I haven't a clue what."

"And another thing, are *all* brains conscious? Is a rat conscious? Is an ant conscious? Where does it stop?"

"Look. I think there are one or two things we can say. We do know that when you are conscious, certain areas of the brain become active which, sort of, shut down when you are unconscious. You can actually see this happening in a MRI scanner."

"Yes – they are going to give me one of those tomorrow."

"I dread to think what they might find inside *your* brain." I said. "Wouldn't it be awful if, one day, they invented a scanner which could actually read the thoughts inside a brain!"

"Is that feasible?"

"Well, in a sense, I don't see why not. Presumably a 'thought' is a certain *pattern* of electrical impulses in the neurons. And a pattern isn't a pattern unless it has some recognisable feature which identifies it. So if you could record all the electrical activity in a brain, you could, presumably and in principle, read its thoughts."

"But would this super brain scanner be able to tell if the brain was *conscious* or not?"

"Good question. No, I don't think so. Even a computer has 'thoughts' in the sense that particular patterns of activity turn up regularly; but we presume that a computer is not actually *conscious*. I suppose it could be that conscious thoughts are merely extra specially complex patterns but somehow that doesn't seem enough to me. No, a conscious thought must be a *process* not merely a *state*."

"What do you mean?"

"I don't know what I mean. Well, perhaps I do. What I mean is that a conscious thought is a dynamic thing; it has to take place *in time*; it doesn't just *exist*. It is *active* not *passive*. When you are conscious, thoughts don't just flit through your mind at random – they have a *purpose*, they *process* information at a high level, they *direct* your future actions, you can *choose* what to think next, you can . . ."

"What you are saying is that conscious thoughts are inextricably bound up with our sense of *free will*."

"Yes! That's it! Exactly!" I exclaimed. "When you are conscious, you can exercise Free Will. When you are unconscious, your brain is just following a program. What a brilliant idea!"

"Modesty forbids . . ." said Alan. "But before you get too excited, it seems to me that you have just replaced one mystery with another, equally, if not even more controversial. What is Free Will? Do rats have it? Do ants have it?"

"Yes – I guess you are right."

"What about your pet theory of suspended reality? Might that shed some light on the issue? Didn't you say once that, in your theory, information from the whole region surrounding a quantum system has to be taken into account in determining what happens to it next? Could it be that the whole brain is somehow affecting the behaviour of individual neurons in it?"

My brain slipped into overdrive.

"Say that again please."

"I don't know what I said, exactly. Some rubbish about the whole brain affecting its own neurons, or something lie that."

"You've done it again!"

"What?"

"Given me an idea."

"What idea?"

"I don't quite know yet. It might take a while to think it through."

"Go ahead. I'm not going anywhere soon." said Alan, reaching for the box of chocolates.

3: What is CONSCIOUSNESS?

Now we come to the most difficult problem of all – what is the nature of that overwhelming sense which (presumably) all humans share with me that, when I am awake at least, I am *conscious* both of myself and of my surroundings?

Any satisfactory answer to this question, if indeed it exists, must also provide answers to the following much more tractable questions:

A) *Are there degrees of consciousness?*

B) *What creatures, if any, other than human beings possess consciousness?*

C) *At what stage in its development does a human child become conscious?*

D) *What are the evolutionary benefits of consciousness?*

E) *Will it ever be possible to attain a proper scientific explanation of consciousness?*

F) *Would such an explanation shed any light on the age-old problem of free-will?*

G) *Will it ever be possible to construct a machine which is conscious?*

But before we even start, it would be as well to review what we know or believe about *un*conscious brains and I shall assume that worms, insects, crustaceans, reptiles and sleeping birds and mammals are all unconscious. (I will discuss the reasons for this list later but, even if you disagree with it at the moment, bear with me. You only have to believe that *some* of the creatures in this list are unconscious for this section to be relevant. Indeed, it would be as well to start by assuming that *all animals* except for non-sleeping humans lack consciousness until we have objective evidence to the contrary.)

The unconscious brain

How do ants walk? How do dragonflies hunt? How do snails find food? How do crabs fight? How do spiders make webs? How does a sleepwalking human make a cup of tea?

Well the first thing to say is that we can build robots to do all of these things so the temptation is to conclude that all these creatures are merely complicated robots, pre-programmed by inheritance or learning to do certain tasks. But is this really true? Are unconscious creatures built on the same principles as a modern laptop?

There are several potential ways to build a robot. You can make a simple line following machine with nothing more than a couple of optical sensors wired up to two electric motors in such a way that if the sensors drift off to the right, the motors cause the robot to turn left and so on. The circuit is pretty simple and an example is shown in *A line-following robot* on page 194. In order to change the behaviour of the device – such as following a white line instead of a black one, all you have to do is change the way it is wired up.

Another way is to use a programmable chip like a PIC. This can be programmed to do exactly the same job but it has the advantage that it can be reprogrammed to do much more complex things such as following the line for 10 seconds, then turning round and going back. The first machine is entirely hard-wired while the second relies on a stored program.

A third way is to wire together hundreds or thousands of microprocessors in such a way that over time, they can be taught to follow the line. Such a device is called a neural network.

The question now arises – are the brains of unconscious creatures also hard-wired, pre-programmed or do they program themselves? Fortunately, we know the answer to this question in the case of the nematode worm *Caenorhabditis elegans*. This unattractive but highly successful species has precisely 302 neurons and we know exactly how they are wired up and what they each do. *C. elegans* is basically little more than a hard-wired line following machine. It is not programmable and has no 'software' and cannot be taught anything. Could it be the case that all unconscious brains are the same? Hard-wired? Non-programmable?

Is the architecture of a spider's brain anything like the architecture of a modern laptop with comparable processing power? Is there the equivalent of a CPU running a program in machine code? Are there distinct areas of a spider's brain dedicated to storing programs and information? I strongly suspect that the answer to all of these question is negative. We know that individual neurons are very like the logic gates that you find in a computer but the way they are wired together in a spider's brain is very different. In a modern PC, each logic gate is connected to relatively few other gates whereas neurons may have dozens if not hundreds of inputs and outputs. In a modern PC, timing is critical because everything has to happen in a precisely choreographed order; in a spider's brain, everything seems to happen at once. A PC is essentially a *serial* machine; it seems likely that a spider's brain works in *parallel*. The conclusion is that the brains of most living creatures are not really like computers at all; they are much more like incredibly complicated line following robots. That is to say, they are hard-wired and essentially non-programmable.

But if this is so, how do such creature *learn*? Honey bees can remember the location of a source of food and communicate that information to the rest of the hive when they return so they must be able to learn. Fish too can adapt their behaviour in response to laboratory situations and can learn migration routes from their elders so if we go along with the idea that these creatures are not conscious, we must grant them some means of storing and recalling learnt information. But does this mean that they have 'memory cells' and possess some sort of software 'code' in which to store information? No – it does not. The hard-wired white line following robot can be converted into a black line following robot by swapping round a couple of connections. It does not need a memory – only the ability to adapt its wiring. And this is a feature that we know organic brains possess. There exist remarkable videos of neurons actively seeking out the right connections to make and when human brains grow old and lose the ability to make new connections in their brains, they also lose their ability to make short and medium term memories too.

It is for this reason that the most popular explanation of how the brains of unconscious creatures like bees and fish work is in terms of a neural network which is partially pre-programmed at birth but which has the capacity to modify itself in the light of experience.

In his book 'The Astonishing Hypothesis' Francis Crick puts forward a powerful case for suggesting that this is the way a human brain works as well. At every stage of its life, the brain is nothing more or less than a huge collection of neurons – about 100 billion of them – connected together is some fabulously complicated way – using up to a quintillion (10^{15}) connections (synapses). There is no software, no secret code. What you see is all there is. In principle, therefore, if you constructed a machine made out of silicon with 100 billion computer chips (each neuron is more like a mini microprocessor than a transistor) all wired exactly like my brain, it would behave exactly like my own brain. If it was connected up to my body in appropriate ways it could regulate my breathing, react to light and touch, it might even cause my body to get up and make a cup of tea and it could probably play Für Elise – but there would be one big difference. It wouldn't talk sense, it wouldn't enjoy classical music, it wouldn't write books on consciousness, it wouldn't recognize my wife as being anyone greatly different from any other lady of my acquaintance – in short, it wouldn't be *conscious;* it wouldn't be ME.

So what is it that changes when I wake up in the morning and my hard-wired robot brain which has quietly been monitoring my breathing all night, digesting the dinner I ate yesterday and generally looking after my body turns into a conscious entity capable of remembering what it did yesterday and deciding what it wants to do today? To answer this we need a theory of the Mind.

Theories of the Mind

The essential dichotomy which lies at the bottom of the mind-body problem goes right back to those two founders of philosophy: Plato and Aristotle. Plato is famous for his theory of forms – the idea that the material world, including our own bodies, is but a shadow of a higher realm of ideal forms in which the immortal soul resides. While we live, our soul is, temporarily, attached to our bodies in the same way that the ideal form of a cube is, temporarily, attached to a child's building brick. This idea, in one form or another has enjoyed enduring popularity down the centuries and still forms the basis of most religious philosophies. It provides a satisfying explanation (to many people at any rate) of the overwhelming sense of self that we all experience whenever we contemplate our own minds as well as conferring other benefits such

distinguished by the varying ways and degrees in which the brain (i.e. the electrical activity of the neurons) and the mind (ie the mental states which emerge from all this activity) interact with each other. For some, mental states are just a way of describing an extremely complex physical state in the same way that the temperature of a gas is just a way of summarising the average behaviour of a large number of gas molecules. Under this view, mental states are peripheral to the important process in the brain. Mental states are nothing more or less than physical states. For every mental state there is a physical state and vice versa; they are two sides of a coin and they develop in parallel.

This is the position adopted by the electronics engineer who insists that a complete understanding of the workings of a laptop computer can, in principle at any rate, be provided by a complete wiring diagram and a list of all the data held in permanent storage. It is not necessary, the engineer would claim, to know the instruction set of the CPU or what language the operating system is written in; if you know how the machine is wired and the state of its memory, you know, in principle if not in practice, everything you need to know to predict what it will do.

The difficulty with this position when it comes to explaining the workings of a conscious brain is that it is not clear what *use* these 'mental states' are to the organism which possesses it. 'Mental states' may be of use to a psychologist faced with the need to help a mentally ill patient but there does not seem to be any reason why the individual needs to be *conscious* of his mental state any more than a gas needs to be conscious of its temperature.

While **parallel Epiphenomenalists** play down the role of mental states, others (who we might label **interactive Epiphenomenalists**) believe that mental states have an essential role to play in determining what the brain does next. We might, for example, imagine that the action of poking a stick into an ants nest provokes a 'mental state' in the colony called 'stress' which itself causes the ants to behave defensively. It is obvious that a single solitary ant cannot possess this state because 'stress' is a property which belongs to the whole colony. Moreover, unlike temperature which is a simple average over a large number of molecules, 'stress' in an ant colony is more a description of the way in which the colony is organised and in this respect it is more like the concept of entropy than temperature. Now there is a respected body of opinion within the physics community which regards the second law of thermodynamics as a truly fundamental law on a par with Newton's laws of gravity or motion and if this were true, then it would be impossible to explain how a petrol engine or a refrigerator works using the laws of motion alone; you would have to use the concept of entropy. In the same way, the student of ant behaviour would be forced to use the concept of 'stress', not just as a short cut but as an essential tool in understanding and predicting the behaviour of the colony.

Advocates of this position face the same difficulty that confronts the Cartesian Dualist: what is the actual mechanism by which 'mental states' in a brain or an ant colony can influence the behaviour of 'physical states'? If we point to the observation that disturbing the ants by poking a stick into the nest causes some of them to release hormones into the air which triggers a defensive reaction in the other ants, then we have already admitted that the supposed mental state called 'stress' is not the actual cause of the reaction; in which case we can, in principle, dispense with it. Similarly, if we discover that 'being in love' is always associated with exceptional activity in a particular cluster of neurons, then we could, in principle, replace that well-known phrase in the literature with its alternative, but much less poetic description.

Finally, we come to the other extreme position which we might describe as **Materialism**. Aristotle was a materialist. The (conscious) mind (or, if you like, the soul) was simply an attribute of a human being in just the same way that being cuboid was simply an attribute of a brick. The modern materialist will not speak of souls or even attributes; he will go further. He will claim that the whole mind/body dichotomy is a red herring; the mind is not just an attribute of the brain, they are, in fact identical. Mental states are physical states – nothing more, nothing less. If pressed with the objection that my *idea* of a football is not an *actual* football, he might be persuaded to say that the distinction between the mind and the brain is a bit like the difference between the software and the hardware in a computer. If pushed hard enough, the die-hard materialist may be forced into one of two corners. He may admit that, ultimately, he does not believe in mental events at all and, if you are really cruel, you may be able to get him to deny his own consciousness. (When not calling him a fool, I would label such a person as a **radical Materialist**.) On the other hand, he might take the view that, since human computers (brains) possess consciousness, other sentient beings must also possess the same quality but in lesser degree. And if you press this argument vigorously enough, you may get him to admit a degree of consciousness to robots, thermostats and even stones (which feel and respond to the force of gravity). This point of view has generously been given the grand name **Panpsychism**.

To summarise what we have said so far, many theories of the mind can be pinned on a column which has spiritualism at the top and radical materialism at the bottom.

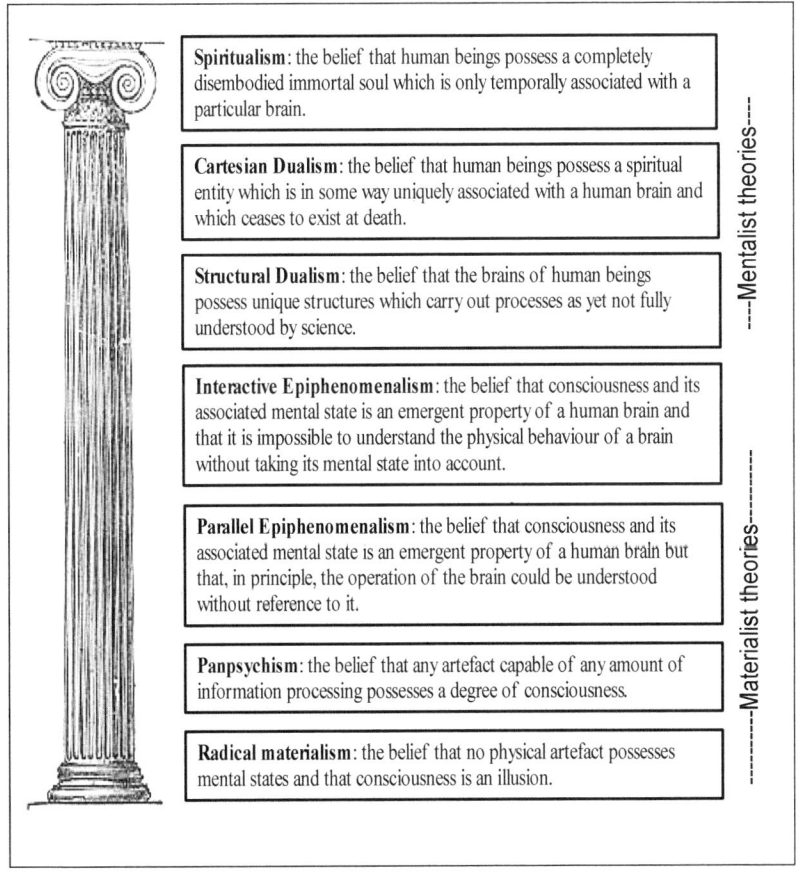

Spiritualism: the belief that human beings possess a completely disembodied immortal soul which is only temporally associated with a particular brain.

Cartesian Dualism: the belief that human beings possess a spiritual entity which is in some way uniquely associated with a human brain and which ceases to exist at death.

Structural Dualism: the belief that the brains of human beings possess unique structures which carry out processes as yet not fully understood by science.

Interactive Epiphenomenalism: the belief that consciousness and its associated mental state is an emergent property of a human brain and that it is impossible to understand the physical behaviour of a brain without taking its mental state into account.

Parallel Epiphenomenalism: the belief that consciousness and its associated mental state is an emergent property of a human brain but that, in principle, the operation of the brain could be understood without reference to it.

Panpsychism: the belief that any artefact capable of any amount of information processing possesses a degree of consciousness.

Radical materialism: the belief that no physical artefact possesses mental states and that consciousness is an illusion.

----Mentalist theories----

----Materialist theories----

The attentive reader will notice that I have sneaked in another 'ism (which I have called Structural Dualism) in the middle and I may as well admit at the outset that this is where I intend to pin my own manifesto.

For centuries, philosophers have argued vehemently for their own favoured positions on this column, usually basing their premises on nothing more than introspection and prejudice; but there is a certain amount of hard evidence which can help us in our quest to solve the mystery of consciousness and it is to this evidence that we turn next.

5 vitally important observations

1) ***Brains are only conscious some of the time.*** Brains can be asleep, drugged or in a coma. If we are to understand what makes a brain conscious, we need to study in detail the difference between the conscious and the unconscious brain.

2) ***Conscious brains can do things which unconscious brains cannot do.*** If this were not the case, there would be no evolutionary need for consciousness. Notwithstanding this argument, we still need to establish exactly what it is that conscious brains can do which other brains cannot.

3) ***Conscious beings possess the ability to memorise extremely complex information (including images and sounds) for long periods of time.*** I shall argue that, although it would appear to be possible for a creature to be conscious without long-term memory, in practice the former is not of the slightest use without the latter.

4) ***Conscious beings report an intense feeling of being aware of their own consciousness.*** In other words, consciousness is a *subjective* quality and, on the face of it, this makes it very difficult to reconcile with the *objective* nature of scientific enquiry.

5) ***Conscious beings also report a strong belief that they can control the future by carrying out certain actions or not as they will.*** The scientific debate on the issue of free will generates such heated responses from both sides that virtually all mind-theorists have completely ignored the subject.

Each of these five observations is telling us something essential about the conscious mind and any theory of the mind which fails to address all these issues is defective in some way. I shall consider each in detail in turn and along the way I shall attempt to shed light on the seven questions listed above on page 52. The first observation, in particular, can, I believe, give us powerful objective evidence as to which creatures are conscious and which are not.

Observation No 1: Brains are only conscious some of the time.

When we are in a deep sleep, we are not conscious. When we are anaesthetized or in a coma we are not conscious. It follows that the mere possession of a brain is not, in itself, sufficient to guarantee consciousness. There are some who maintain that consciousness will always emerge as a natural by-product whenever a system reaches a certain level of complexity but it is clear that complexity on its own is not a sufficient criterion – it all depends on how that complexity is organised. It is also well-known that the brain organises itself differently in different states of sleep and wakefulness. In particular there is a very interesting state called REM sleep during which the brain appears to be just as active as when it is fully awake and it is in this state that subjects, when woken, often report that they have been dreaming. (For more detail on this, see: ***Evidence from the Encephalograph*** on page 195)

The question which interests me is this. *Are we conscious during REM sleep*? I hear a chorus of replies – but the shouts of "Of course we are!" are almost equalled in volume by those who hold the opposite view. To those of you who responded in the affirmative I ask why it is that, although you entered REM sleep 5 or 6 times last night, you are totally unaware of that fact. Surely if you were conscious during those times, you must have been conscious that you were conscious? And to those of you who replied "of course I wasn't conscious – I was asleep!" I ask you to recall at least one occasion when you had a vivid dream. Were you not conscious of that dream? How can you possibly dream if you are not conscious of what you are dreaming?

We are in danger of tying ourselves in knots here but the point is a really important one. All the physiological signs indicate that we are indeed as conscious during REM sleep as we are when we are awake; the difference being that certain functions (like the ability to move our muscles at will and to respond to stimuli) are deliberately suppressed (probably to save ourselves from self-harm). If, at the same time, our memory circuits also suppressed, that would explain why we are so rarely able to remember our dreams and why we cannot recall being conscious during the night.

It would seem to me that, in this instance, we must accept the physiological evidence from the EEG machine over our subjective experience. During REM sleep the brain seems to be working overtime. We don't know what it is doing but it certainly appears to be doing the same sort of things that it does when we are awake. I know that I was conscious yesterday evening because I remember watching the 10 o'clock news. The fact that I cannot remember what I was thinking 2 hours later is not proof that I was not conscious then – only that I cannot remember the conscious thoughts which I had at that time.

If, therefore, we accept that we *can* be conscious during REM sleep (but never during non-REM sleep), it would seem to be highly likely that *other animals which also show similar patterns of EEG activity during sleep are also conscious*.

Many, if not all, animals sleep – including invertebrate species – but nobody is quite clear why sleep is necessary; indeed, it would appear to be a rather risky option. Sleep is not a problem for predators at the top of the food chain or animals which can hide themselves effectively but it can put other animals in serious danger. Some animals (e.g. monkeys) live in social groups so that some members can sleep while others keep watch; others live in large herds for much the same reason. Marine mammals such as dolphins and whales have come up with another solution: they sleep with only half of their brains at one time! Seals can do both. They can sleep with one half of their brain while out at sea, but sleep with both halves while safe on land. Many birds also employ unihemispheric sleep and it is said that migratory birds can sleep on the wing. This has not been conclusively proved for obvious reasons but I see no reason to doubt it because it is only the areas of the brain associated with consciousness which shut down during deep sleep. If sleepwalkers can get up and make a cup of tea without being conscious of so doing, I see no reason why a swallow cannot fly in its sleep. (Sleepwalkers are not 'acting out their dreams' as used to be thought; their EEG patterns are those of non-REM sleep, not REM sleep.).

Now it is a fascinating fact that while most animals sleep, only mammals and birds exhibit REM / non-REM cycles of sleep in greater of lesser degree.[6]

Reptiles need sleep as well as birds and mammals but their EEG

---

6   There is some evidence that cuttlefish also exhibit REM sleep and the same may be true of octopuses and squids.

waves do not show any evidence of a REM like phase. This seems to suggest that sleep in reptiles is more a way of passing the time and giving the body a rest than anything to do with the demands of the brain. Fish too sleep, but their brain activity is difficult to record. Some form of sleep appears to be necessary even for insects and crustaceans but although depriving these creatures of sleep impairs their ability to learn, measurements on their nervous systems during sleep shows no evidence of a REM type phase.

If, then, we go along with the idea that REM sleep is indicative of consciousness, then all mammals and birds are conscious in some degree but reptiles are not. If this is true, it raises an interesting question with regard to the evolutionary development of these families. The common ancestors of these groups are small lizard-like creatures called amniotes which lived in the late carboniferous period some 300 million years ago. Their eggs were encased in a sack containing amniotic fluid and this enabled them to reproduce on dry land without having to return to water. This evolutionary branch quickly divided into two, the synapsids (which developed into mammals) and the sauropsids (which became reptiles, dinosaurs and birds). Now since, according to my thesis, reptiles are not conscious, this would seem to imply that consciousness has evolved separately in mammals, birds and possibly some cephalopods. (An alternative possibility is that reptiles have lost the capacity.) It would also appear that there is a strong correlation between blood temperature and consciousness. Indeed, judging by the fact that the human brain uses 10 times as much oxygen per kilogram as the rest of the body, I would go so far as to suggest that being warm-bloodied is a necessary condition for consciousness. (The fascinating question as to whether any of the dinosaurs were conscious will probably turn on whether or not they were warm-bloodied.)

Evidence from anaesthetics

Before we leave this highly instructive topic, is there anything to be learnt about consciousness through studies of anaesthetics? When we go into hospital for a major operation under general anaesthetic, we do so under the expectation that, however much the surgeon cuts, slices and stitches up our bodies, we will feel no pain at the time and will emerge from the theatre with no memories of the experience whatsoever. We explain this to ourselves by saying that, during the operation we are simply unconscious. But how do we know this? I have argued that, during REM sleep (and specifically while dreaming) the brain is in a conscious state – but we are unaware of the fact because when we wake up in the morning we usually have no memories of the experience. Could it be the same during open heart surgery? Do we, in fact, feel every incision of the knife, every snip of the scissors, every prick of the needle at the time, but simply have no memories of the ghastly experience when we wake up? What a terrifying thought! How do we know that we are unconscious during non-REM sleep? How do we know that the patient in a coma is unconscious?

The short answer is that we don't. But we must not ignore what little objective evidence there is. EEG studies show us that the brain can exist in one of several recognisably different states of activity. One of those states (the rapid, random electrical oscillations associated with being awake or in REM sleep) is definitely connected with the subjective experience of consciousness. The other states are associated with periods of which the subjects later report having had no conscious experience. Why should we doubt them?

Consciousness and pain

There is another way in which we can judge whether a person is or is not conscious and that is by studying their response to painful stimuli. Of course, even when asleep, the brain is constantly monitoring its surroundings and carrying out primitive remedial actions in response to stimuli. If you shine a light on a sleeping person, they will probably turn over and bury their head under the pillow; if you remove the bedclothes they will probably curl up to keep warm; if you make an unusual noise like the sound of breaking glass, they will probably wake up. None of these responses requires action from the conscious parts of the brain. Even if you inflict pain, for example by pricking them with a needle, the sound sleeper will probably react by merely withdrawing the limb. What they will not do is sit up and say "Ouch! that hurt! What did you do that for?". (Even if the subject is enjoying REM sleep at the time and who is therefore, by my theory, conscious will probably not sit up and complain either because, for some reason, subjects in REM sleep are even more difficult to wake up than subjects in deep sleep. The difference comes later when you ask them what happened during the night. The deep sleeper will have no recollection of the event at all but the REM sleeper will say "It's funny you should ask about that. I had this curious dream in which I was in a jousting tournament and I got stabbed in the arm ...")

It is now accepted that pain has evolved because it has survival value. If you accidentally put your hand on a hot surface, the pain you experience will rapidly cause you to take appropriate action to withdraw the hand from the source of heat. Notice that this is not the same as the familiar knee-jerk reflex which is not under the control of the brain; your reaction to the hot surface requires a response from much higher up the nervous system. In fact, it would appear that pain goes, as it were, right to the very top and that, in order for it to be of any use as a survival mechanism, the subject has to be conscious in order to experience and therefore to react to pain. It follows therefore that, with the sole exception of subjects in REM sleep, if the subject fails to produce any of the usual responses to painful stimuli that a conscious person would produce, the subject must be unconscious. Sleepwalkers are pretty oblivious to pain and can do themselves serious harm. We are therefore right to conclude that they are unconscious – a conclusion supported by evidence from their EEG patterns.

If we apply the same test to animals, it is immediately apparent that all mammals show exactly the same difference in response to painful stimuli when they are awake and when they are asleep as humans do. Cats and dogs, rats and mice can be anaesthetised using exactly the same drugs as are used on humans. There is little room for doubt. All mammals can experience pain and therefore all mammals are conscious (some of the time and in some degree). Although the evidence is more difficult to obtain, birds too can be anaesthetised but it is less clear how their responses to painful stimuli change under these circumstances. The issue as regards cephalopods is inconclusive at present. Although I imagine that fish can also be 'put to sleep' using drugs, I doubt whether we need to go any further than appealing to the analgesic (pain-numbing) rather than the anaesthetic properties of the drug to explain any changes in the behaviour of the fish. Other animals such as insects and crustaceans show little evidence that their response to traumatic stimuli can be changed reversibly by anaesthetic drugs so I think we can be reasonably confident that they do not experience pain and are therefore not conscious in any sense which implies a degree of similarity with what humans describe as consciousness.

Consciousness in human children

It is time now to raise the extremely emotive question of when, in its development from embryo to child, does a human being become conscious. First a few facts: the brain starts to develop after about 8 weeks gestation; by 22 weeks, the foetus shows certain primitive reflex actions and a few weeks later the central nervous system is fairly well developed. By 32 weeks the foetus can probably see and hear, smell and touch (though obviously *what* it can see etc. is rather limited!). At birth, the infant brain is fully developed in so far as all the relevant structures are there, its EEG patterns are much the same as for an adult and its responses to painful stimuli can be modified by anaesthetics – but is it *conscious*?

There can be little doubt that the answer is yes and that the faculty of consciousness develops rapidly during the last 10 weeks of gestation. However, lacking any experience of world outside the womb, it cannot be conscious of very much and whether or not it is in any sense conscious of *itself* is a question to which I shall return.

Observation No 2: Conscious brains can do things which unconscious brains cannot do.

I think that most of us (the panpsychist excepted) would agree that computers are not conscious. But the list of things which computers and computer-controlled machines can do is impressive and likely to become even more so. Computers can beat almost anyone at chess; they can diagnose illnesses; prove mathematical theorems; build cars; guide missiles to a target; explore distant planets etc. etc. etc. But no computer has (yet) invented a new joke, written a decent poem or composed a symphony. These examples seem to suggest that what a computer lacks is the ability to *imagine* and *create* new things which have never been imagined or created before. Indeed, if I were asked to adjudicate in a Turing test between a computer and a human being, that is what I would ask the computer/human to do – create something. Of course, there are many human beings who would not pass this test (myself included) but that is not the point. If, as a result of my request, the terminal printed out a really novel joke or the score of a brand new symphony, I would conclude that the being behind the screen was human.

Having said that, I do not entirely rule out the possibility that a computer made of wires and silicon could pass this test or that a machine which passed this test was actually conscious. All I am saying is that, with our current level of technology, there are still certain things which conscious human beings can do which computers cannot.

What then of supposedly unconscious animals like insects and crustaceans? Are there things which mammals and birds can do which these creatures cannot? Do mammals and birds show any evidence of imagination or creativity? Are there problems which insects just cannot solve? What about fish? Do they fit into the pattern?

I am no expert on these matters but on the whole I think that many mammals and birds do indeed show evidence of imagination and creativity which is completely lacking in other lower animals. One of my favourite stories (which is probably apocryphal) is that of the chimpanzee who was faced with the problem of retrieving an apple floating just out of reach in a bucket half filled with water. The experimenters wanted to see if the chimp would realize that, by dropping a brick in the water, the water level would rise sufficiently to allow him to reach the apple. They were astonished when the chimp ignored the proffered brick and instead reached the apple by peeing into the bucket! Ravens can solve similar problems but can you imagine a spider or even an octopus (supposedly the most intelligent cephalopod) doing this?

Now it might be argued that a spider, faced with a garden shed, a spade and the branch of an overhanging tree, wishing to build a web, is faced with a serious problem which requires imagination and creativity in its solution – but you would be wrong. It would be a relatively trivial matter to program a computer-controlled robot to do this task. In fact a surprising amount of animal activity is, indeed, hard-wired. Consider the act of building a bird's nest or the ability of a bee to construct a honeycomb. These animals do not have to think about what they are doing or remember what they have done; they are just programmed to do the job. It is a relatively easy task to program a simple computer to build a nest or a web. What is more, the robot will show exactly the same ridiculous behaviour as the animal if it is faced with a situation which is new – for example, building a nest on a twig that looks strong enough but bends when the nest is built, or getting a spider to build a web in zero gravity. Douglas Hofstadter describes the behaviour of the

Sphex wasp which repeatedly checks its nest over and over again whenever the experimenter disturbs its routine by moving its catch away from the nest. It is clear that the wasp does not have the *imagination* to modify its routine to cope with the new situation.

So what kinds of problems do mammals and birds have to solve that a computer would find really difficult? This is not an easy question to answer but since, presumably, consciousness has evolved because it gives its owners a competitive edge, the answer must involve either finding food, finding a mate or rearing young. Now many mammals (including whales) and birds travel vast distances in search of abundant food or good nesting sites, so having a conscious brain may be of use to them in performing the necessary navigation. It is certainly difficult to explain how a pigeon can find its way home even when it is transported hundreds of miles away from its roost; how elephants can return to a water hole after many years of absence and how many birds return year after year to the same nesting sites. How do they do this? It is tempting to suggest that they have a conscious awareness of their surroundings.

On the other hand, monarch butterflies migrate thousands of miles, cruise missiles do exactly what homing pigeons do and even the humble limpet returns to the same spot on its rock after a days foraging so this is no proof that mammals and birds are conscious. And in any case, this sort of feat does not seem to require imagination or creativity.

Perhaps imagination is required when an animal is required to change its normal habits as a result of habitat loss or climate change. I don't think anyone will argue with the thesis that the success of mankind in dominating this planet is entirely due to his unique ability to adapt his behaviour according to the circumstances he finds himself in and that this adaptability springs from his ability to use conscious reasoning but it would be difficult to extend this argument to other putatively conscious animals.

I think the real answer must lie elsewhere. The one feature that really distinguishes mammals and birds from fish, crustaceans and insects is their ability to *recognise each other as individuals*. This is obviously true of the apes and I believe it to be true of elephants, whales and dolphins too. All social animals such as lions and wolves need this skill as do all birds who mate for life. The hypothesis is more difficult to prove in the case of smaller mammals and birds but any creature who suckles its young or feeds a chick would be well advised to be able to recognise its own offspring. (I am afraid that the poor willow warbler who feeds a cuckoo chick twice its size cannot be credited with much imagination and it is in all likelihood not conscious of what it is doing!) I think it extremely unlikely that any insects or crustaceans have the ability to recognise other members of the species as individuals and I would be very interested to know if any fish (such as sharks, which definitely hunt in packs like wolves) can do it. My gut feeling is that sharks, like bees or ants, behave cooperatively in the hunt by instinct and whereas I can imagine a wolf thinking to itself (in wolvish) 'loppy-tail is a fast runner, I will leave the chase to him and go round the back here to cut off the deer's retreat', I can't see a shark thinking like that.

My conclusion is that the possession of a conscious mind permits creatures to have *personal relationships* and it is this, more than anything else which gives mammals and birds their evolutionary advantage.

Observation No 3: Conscious beings possess the ability to memorise extremely complex information (including images and sounds) for long periods of time.

It is almost as difficult to understand how our memory works as it is to understand consciousness. It is pretty well established that the human brain does not store memories in the same way that a computer stores information. There is no single collection of neurons in your brain that holds your credit card PIN number. The metaphor of a hologram is probably more helpful or even that of a fractal algorithm which somehow conjures up an image of a fern. It should also be remembered that humans have at least three different kinds of memory and it is probable that different methods are used to store information in each case. Short-term memory – the memory that you use to write down a telephone number a few minutes after you have been told it and the kind of memory that I always appear to use whenever I am told the name of a new acquaintance! – is probably dynamic in the sense that it requires the continuous firing of certain neurons and is almost instantly forgotten. Medium term memory must use a slightly more permanent method of storage but true long term memories are probably held as a result of almost permanent changes to the way that the neurons in your brain are connected together.

Now it is often said that 'elephants never forget'. I don't suppose that elephants are really any less likely to forget things as we humans are but what is indubitable is that they can remember things for a long time. I have already mentioned their ability to remember the location of a water hole last visited many years ago and the sight of a young elephant trumpeting over the bones of his mother killed by poachers months before is poignant testimony to their ability to remember past events. Dog handlers will recount stories of impressive feats of memory by their pets and penguins can recognise their mates after months of separation at sea. Recent research suggests that dolphins can remember the calls of individuals which they last met as long as two decades ago. Even rats (who have become familiar with several different mazes) can remember where the bait was placed last time they ran the maze for at least a week.

Nothing here suggests any necessary link with consciousness, though, and it may indeed be the case that the two phenomena are either entirely distinct or more probably two products of the same feature of the brain but it does seem at least plausible that consciousness and long-term memory are closely linked and that you cannot have one without the other. If, as I have suggested above, the evolutionary advantage of consciousness is to enable the creature to engage in long-term personal relationships with other members of the species then there is no point in being conscious if you can't remember what you were once conscious of.

I would be interested to know of any evidence that any of the lower orders of animal possess a long-term memory. Apparently French angelfish form lifelong bonds but this is very rare and can probably be explained without any reference to conscious memory but simply in terms of adaptive behaviour. I would be extremely surprised if, after a period of separation, two angelfish would necessarily resume their former cooperative relationship with each other.

Since, as I have argued above, consciousness also appears to be intimately associated with long-term memory, I would also like to suggest that the subjective experience of consciousness necessarily involves the presence of sensory inputs which are interpreted by the conscious brain *in terms of its past experiences*. The implications of this are quite profound. I am suggesting that the new-born infant, although in possession of a fully functioning human brain, can only really be said to be conscious in a technical sense. Since it has no memories of past experiences with which it can interpret the sudden change in sights, sounds and smells which suddenly assault its senses, it cannot truly be said that it is *conscious* of them. Gradually, as the days and weeks go by, the infant learns to make sense of the the external world and eventually becomes conscious of it and of its own place within it.

Interim conclusions

We are now in a position to answer the question 'what creatures other than ourselves are conscious?' with some degree of confidence.

Previous attempts to define consciousness have often concentrated on the ability of conscious beings to sense their surroundings and to react to changes. Such theories have sometimes led to the suggestion that consciousness is not just an attribute of humans but is possessed by all living creatures in a certain degree. Indeed, some philosophers have even credited the property to rocks and stones which are 'conscious' of the Earth's gravity if nothing else. It is my contention, developed in the next few pages, that consciousness is restricted to beings which can exhibit patterns in the brain of such delicacy and complexity as to influence the behaviour of matter through some unknown non-classical process such as quantum collapse – a process which enables the creature to recognise itself as an individual and to plan for the future. This rules out rocks and stones. It also probably rules out all plants and lower forms of animal life on the basis that their nervous systems are insufficiently complex. Digital computers can be programmed to perform remarkably intelligent tasks but even our most powerful computers do not form personal relationships with other computers; nor can they use non-classical processes to exercise their free will so no current computer, not even the proposed quantum computers of the foreseeable future , can be conscious.

So is my pet goldfish/octopus/dog/wife conscious?

Obviously we are not currently in a position to ask whether or not these creatures employ non-classical processes in the brain so instead we must rely on the answers to the following four questions.

1. *Is its brain sufficiently complex*? (Goldfish: probably not; Octopus: very probably; Dog: almost certainly; Wife: certainly)

2. *Does its brain show typical changes in patterns of activity when the creature is awake or asleep*? (Goldfish: probably not; Octopus: unknown; Dog: certainly; Wife: certainly)

3. *It is capable of forming long-term relations with other members of the same species*? (Goldfish: definitely not; Octopus: probably not; Dog: certainly; Wife; certainly.)

4. *Does its behaviour show evidence that it plans what it is going to do before it does it*? (Goldfish: certainly not; Octopus: very possibly; Dog: certainly; Wife: you must be kidding!)

Joking apart, it seems to me that the outcome of this test indicates that the only conscious creatures on this planet are humans and other social mammals (including whales and dolphins, of course) and some of the more intelligent birds and possibly octopuses, cuttlefish and squids. The implication of all this is that you need feel no guilt about stepping on a beetle or throwing a crab into a pot of boiling water. On the other hand, we do need to think about how we treat our farm animals and bird life.

Since consciousness depends just as much on long-term memories of past experiences as it does on non-classical processes going on in the brain, we can also conclude that some creatures with limited memories of past experiences (including new-born infants) are, in a very important sense, *less conscious* than ourselves. Again, I believe that this conclusion could have important implications for debates about abortion and other ethical issues concerning life and death.

Towards a definition of consciousness

So far we have discussed a number of ways in which we can objectively identify the probable occurrence of consciousness in creatures other than ourselves; but we have not begun to address the question of what consciousness actually *is* or how the subjective experience of being conscious actually comes about. Are there any pointers at all to where we should start looking for the extra ingredient which is necessary to convert an unconscious brain into a conscious one? Is it just a question of organization (as the epiphenomenalists would have it)? Or are we going to have to take paranormal phenomena like mind-reading and telekinesis seriously in our search for the key to consciousness (as the spiritualists would have us believe)?

No open-minded scientist should rule out either possibility, but there is another alternative and that is that there is some perfectly rational physical process going on in the conscious brain of which we currently have no understanding whatsoever. Lightning was considered an act of God until Franklin showed that it was just a form of electricity; Heat was considered to be a type of fluid until Joule showed than it was a form of energy; Gravity was thought to be something called a 'force' which propagated instantaneously over vast distances until Einstein showed that it was due to the curvature of space-time. And if astronomers can trump all the evidence patiently collected by our particle physicists using their vast and expensive accelerators by proposing that most of the universe is composed of something which nobody has ever seen just on the basis that a few incredibly distant galaxies are a bit dimmer than they ought to be, surely I can be forgiven for proposing that, since the phenomenon of consciousness cannot be adequately explained by existing scientific theories, it must be because it relies on some property of nature which we have yet to discover.

Now it is tempting to suggest that, since there are in fact two things which we fundamentally do not understand – namely consciousness and quantum theory – the solution to one riddle may lie in the other. In fact, it has been said that this is the only argument in favour of a quantum theory of the mind but I have to disagree. I believe that there is at least circumstantial evidence to support this view. When two particles enter an entangled state they could be said to have, temporarily, become one entity with properties quite different from the two particles which they eventually turn back into. Perhaps bits of our brains have the capacity to enter quantum states with similar non-classical properties. Perhaps our conscious brains are quantum, not classical computers and it is this which gives them the power to dream up new jokes, invent new mathematics and compose symphonies. Although this idea may seem far-fetched, it would, at least, explain how the same brain can sometimes be conscious and sometimes unconscious. When it is conscious it is operating in quantum mode; when unconscious it is operating in classical mode.

One minor objection to this idea is that the process of creativity (which I have assumed is the hall-mark of a conscious creature) often seems to take place in our unconscious minds. I am told that Mozart would wake up one morning with a complete symphony in his head; and we all know that when we have a particularly knotty problem to solve, the best thing is to go and do something else and often the answer will pop into our heads by itself. I do not regard this objection as serious. I have not claimed that conscious experiences *necessarily* accompany quantum processes in the brain and I deem it perfectly possible that the latter can occur without the former. It seems entirely likely that there are degrees of consciousness and that, even while we sleep, there may be parts of the brain which continue to operate in quantum mode (e.g. during the REM phases). Since, as I have argued above, consciousness also appears to be intimately associated with long-term memory, I would also like to suggest that the subjective experience of consciousness is not just a product of quantum processes in the brain but also necessarily involves the presence of sensory inputs which are interpreted by the conscious brain in terms of its past experiences. To put it another way, consciousness without past experiences is not true consciousness at all.

Putting all of these ideas together I would like to suggest the following tentative hypothesis: *consciousness is a result of a particular way in which a brain uses non-classical operations to process incoming sensory information in the light of past experiences to generate novel solutions to problems posed by its environment.*

This places the study of consciousness on a firm scientific basis. If we eventually do discover a structure or process in the human brain which is uniquely correlated with conscious experiences (even if it turns out to be a perfectly classical one) we will have learned a great deal about what causes consciousness and we should be able to use this knowledge to predict the behaviour of other creatures which do, or do not possess the same structures.

But even if we were to reach this stage of enlightenment, would we be in a position to *explain* our *subjective experience* of consciousness, our sense of *being*, our sense of our own *uniqueness*? I am afraid that the honest answer to this is no. But then it is not the job of a scientific theory to explain *everything*. Newton's law of gravity can be used to explain why planets move in ellipses and it can be used to predict the motions of the planets with uncanny accuracy; but it says precisely nothing about why massive bodies attract each other in the first place. Einstein's General Theory of Relativity could be said to answer this question but only by posing another – namely, why do massive objects bend space-time?

Even if we knew so much about the unique structures and processes that go on in a conscious brain that we could construct conscious machines out of string and sealing-wax, we still would not know *why* our machine was conscious. It follows that we human beings will *never* understand the workings of our conscious minds in any scientific sense. The best we can hope for is that an increased understanding of the physical laws which govern matter will give us insight into the way in which the conscious brain interprets sensory information in the light of remembered experiences.

So the only scientific question which is worth asking at this stage is the following: *what are the unique structures and/or processes in the human brain which are uniquely correlated with consciousness and do these structures or processes really carry out non-classical (e.g. quantum) computations or not?* In other words, is the conscious brain a non-classical (e.g. quantum) computer or is it merely a very complex classical one?

The great majority of neurophysiologists work on the assumption that it is the latter because all the micro-structures and processes in the brain which have been observed so far can all be explained adequately in classical terms. But there are a few intrepid scientists who are prepared to think the unthinkable, the most prominent of whom is Sir Roger Penrose who has made a powerful case for the non-classical nature of the brain and who has even identified some parts of the brain where these non-classical processes might occur. In his book 'The Emperor's New Mind' Penrose uses Gödel's theorem to 'prove' that the human brain cannot be a classical computer. Now it must be conceded that many if not most scientists do not accept that Gödel's theorem has anything to do with the human brain but, even if his logic can be questioned, his conclusion may still be correct. Conscious human brains do appear to be able to perform feats which it is extremely difficult to imagine computers performing *however powerful they may become* in the future. One of these feats is the ability to analyse itself by being *self-aware*.

Observation No 4: Conscious beings report an intense feeling of being aware of their own consciousness.

When philosophers and scientists talk about consciousness, one word crops up more than any other and it is: *awareness*. But if any single word has generated more misunderstanding and confusion in the subject than this one, I know it not. It is high time that we abandoned the use of this word and adopted a scale of (semi-technical) terms to describe the various levels of awareness that different systems, both organic and inorganic, display. Might I suggest the following list?

> *Susceptibility*: I use this term is the basic sense to describe anything that is susceptible to an external influence. For example, the Moon could be said to be aware of the Sun because it is *susceptible* to the Sun's gravity. A badly built house could be said to be aware of earthquakes because it is *susceptible* to being shaken.

> *Irritability*: this term is used by biologists to describe the awareness that plants and primitive animals show when they respond to changes in their environment. For example: plants may open their flowers when a light is shone on them; woodlice retreat into the darkness; bees swarm when their hive is disturbed. I think we can also extend the use of the term to describe the behaviour of any man-made device which has sensors and which uses the information from these to control its actions. Such machines would range from a simple thermostat to Google's driverless car

> *Sentience*: this is defined as having the power of *conscious* perception through the senses. Just look around you and be aware of your surroundings. That is *sentience*.

> *Self-awareness*: Look at your hands; touch your face; recognise that in an important sense, these things are different from the other objects around you; they belong to *you*; they are under your control. That is *self-awareness*.[7]

---

7   Note that my use of the term 'self-awareness' only refers to my bodily self. I do not wish to imply the existence of any sort of mental 'self'. A creature with a well-developed sense of self awareness should be able to pass the 'mirror test'. The fact that dogs do not pass this test does not necessarily mean that they are not self-aware, however.

Auto-consciousness: close your eyes and try to think of your own consciousness. If you can do that, you can be said to be aware of your own consciousness i.e. you are *auto-conscious*.

It is probably clear from what I have said earlier that I regard insects, crustaceans and other primitive animals with rudimentary brains as irritable, not sentient (but I concede, this is far from proven).

When it comes to categorizing conscious beings, the situation is more difficult. There is good evidence that primates, elephants and grey parrots are capable of self-awareness (paint a spot on their forehead while they are asleep and then put them in front of a mirror.) but other creatures may also be self-aware without passing this particular test. At what point does a human infant become self-aware? That is an important and interesting question.

Another interesting question is this: are all conscious beings auto-conscious? Or to put it another way, is it possible to be sentient or even self-aware without being aware of your own consciousness? Just because we humans find this difficult, it does not mean that it is logically impossible. It is fashionable these days to play down the difference between human beings and the higher primates to the extent that the latter are now afforded certain legal rights in some countries. But if it could be shown that, while higher primates are self-aware, only humans are auto-conscious, would this change matters?

But there is another dimension to conscious awareness which I have not mentioned yet and that is our awareness of the passage of time and of our own immediate past. I think we should call this **temporal-awareness**. In some ways, I think this is the most important aspect of our conscious awareness and is the reason why I included the phrase 'in the light of past experiences' in my definition of consciousness. To my mind, a new-born baby cannot be said to be truly conscious because its past experiences are so limited. To be fully conscious you have to be sentient, self-aware, auto-conscious *and* temporally-aware.

But this neat classification of degrees of awareness has its problems. In what state of awareness is the dreamer or the hallucinating drug addict? EEG test appear to show that both are conscious in the sense that their brains seem to be doing the same sort of things that brains do when they are awake – but they cannot be said to be sentient let alone self-aware or temporally-aware. The brain appears to be doing its quantum thing without reference to either current sensory data or past memories. We could usefully describe this paradoxical state as being *conscious* but *not sentient.*

This is all very well but inventing pretty definitions does not prove anything. I agree. But it does help to sort out what is important from what is not. The big question is – does our ability to be sentient, self-aware, auto-conscious and/or temporally-aware shed any light on the processes that are going on in our brains? Would it be possible to design a machine to be any of these things? The epiphenomenalist would argue that all these things are possible if the system in question is complex enough and designed in the right way. My feeling is that there is something qualitatively different between sentience and irritability which cannot be bridged by a mere increase in complexity or subtlety of programming. But in truth I cannot refute his position convincingly. I just don't see how a classical computer could be aware of itself.

But I can already hear the epiphenomenalist's triumphant reply: 'I don't see how a quantum computer could be aware of itself either!'

So lets look at this objection more closely. Is there anything in quantum mechanics which could lead to self-awareness or any of those other types of awareness associated with consciousness? I believe there is.

One of the most important aspects of quantum theory is its essential non-locality. In classical physics, all causes are local. The moon moves in such a way because it is responding to local changes in the curvature of space-time; a photographic plate records and image because it is hit by an electromagnetic wave of energy; a billiard ball changes its direction of motion because it is struck by another ball etc. etc.. But in the quantum world things seem to happen either without a cause (e.g. the decay of a radioactive atom) or because a measurement is made somewhere else (as in experiments on entangled particles). Now whatever interpretation you adopt regarding quantum theory, you have to conclude that it is not possible to treat any quantum system in isolation; it is always fundamentally connected to its environment. If we adopt the hypothesis that the conscious brain operates on quantum not classical principles, then we would expect that conscious processes in the brain would also be essentially non-local. In short, we might expect conscious thoughts to take place in the whole brain (or large parts of it) rather than separately in collections of individual neurons.

Now it is a curious fact about our conscious brains that we can only think of one thought at a time. Consider what happens when you are walking along a road, deep in conversation with a friend. The time comes when you have to cross the road. The conversation stops. You assess the traffic and cross when the road is clear. The conversation resumes. Why did you not continue the conversation while assessing the traffic? The answer is that both processes required the whole brain to compute. It just is not possible to think two different thoughts at once. On the other hand, a computer controlled robot would have no difficulty at all in carrying out both processes simultaneously (It could do this either by employing two microprocessors independently or by time-sharing, it doesn't matter which.) but this is not, in general, possible for a human. (It is conceivable that patients with a surgically 'split' brain could think two different thoughts at once and psychological studies of such patients could shed much light on the potential ability of one half of the brain to communicate with the other by non-local quantum processes.)

Now it is this single-mindedness of the human brain which is responsible, I believe, for the overwhelming sense that we are each a unique individual and I find it highly suggestive that this single-mindedness is a necessary consequence of the hypothesis that the brain is a quantum computer. In short, it is a necessary condition for a conscious quantum brain to be in some important sense in communication with the whole of itself. And if this isn't a good definition of self-awareness, I don't know what is.

And there is another reason why it is extremely tempting to believe that the human brain is a quantum not a classical computer and that is to do with the existence or otherwise of *free will*.

Observation No 5: Conscious beings report a strong belief that they can control the future by carrying out certain actions or not as they will.

If you take the view that the brain is merely a complicated (but conventional) computer and a human being just a robot, blindly obeying the laws of physics and totally incapable of making real decisions or creating anything new, then you must deny the existence of Free Will..

Even if the robot is, in your view, 'conscious' and possesses a sense of self awareness, it makes no difference; it cannot possess Free Will.

This is a perfectly tenable position but it is, to me, unacceptable. It is, of course, quite impossible to prove that my decision to marry my wife was not just predetermined by the laws of physics (and possibly also the laws of chance) but I really want to believe that there is an 'I' inside me who had some choice in the matter – otherwise, what is the point?

Actually, as an objective realist, there is no real reason why I should expect there to be a point in anything and so this is not, in fact, the reason why I want to believe in Free Will. There are good empirical reasons for believing in Free Will, the main one being the fact that humans, generally, do what they say they are going to do. If the world was predestined by the immutable laws of Physics or chance, then a murderer would be just as likely to utter the words 'I am going to kiss you' as 'I am going to kill you' before plunging the knife into his girl-friend. In a predestined world law-abiding people would be just as likely to end up in jail as murderers. In a predestined world a dog would as soon chase a bird as a ball. In a predestined world a squirrel that had buried a nut in a forest would be no more likely to find it again than any other squirrel in the forest.

Now it is no coincidence that all my counter-examples given in the previous paragraph necessarily involve a creature which seems to know what it wants to do before it does it. A predestined world may evolve plants and humble animals but it could not contain a human or a chimpanzee because these creatures can be trained to tell us what they are going to do before they do it. Other creatures such as dogs and squirrels cannot actually tell us their intentions, but few dog owners, faced with an excited, tail-wagging dog with a stick in its mouth, would deny that their pet possessed the ability to decide for itself what it wanted to do. And the behaviour of squirrels is incomprehensible without assuming that they, too, have a degree of self awareness and an understanding of the relationship between themselves and their environment which amounts to a degree of consciousness.

In fact, the conclusion seems inescapable that it is only conscious minds which can exercise Free Will – and minds which can exercise a degree of Free Will must be, in similar degree, conscious.

But when translated into scientific language using the 'mind is a brain pattern' axiom, this same sentence becomes highly controversial and possibly even contradictory. It becomes: *Certain patterns of activity in subsystems called brains can affect the laws of Physics in such a way that certain possible outcomes become more likely than before*. Or, more succinctly: *Minds can overrule the laws of Physics*.

Immediately we are plunged into a circular argument which seems to lead to an impossible contradiction – if minds are just brain patterns which obey the laws of Physics, how can such patterns break the laws of Physics?

The classical explanation

Newton's laws of motion seem to imply that the future is completely pre-determined. Pierre Laplace famously said in his book 'A philosophical Essay on Probabilities':

> *An intellect which at a certain moment would know all forces that set nature in motion, and all positions of all items of which nature is composed, if this intellect were also vast enough to submit these data to analysis, it would embrace in a single formula the movements of the greatest bodies of the universe and those of the tiniest atom; for such an intellect nothing would be uncertain and the future just like the past would be present before its eyes.*

At a stroke, Free Will was abolished.

Worse was to follow. Henri Poincaré proved a theorem which seemed to suggest that in an infinite universe, events would repeat themselves over and over and over again. Not surprisingly, every effort was made to find a way out of this horrifying prospect but it proved to be no easy task.

One chink of light was provided by Poincaré himself who showed that certain systems could be extremely sensitive to initial conditions and that, while Laplace's supreme intellect could, in principle, perform the feat described, it would be completely impossible to determine the positions and velocities of all the balls in a lottery drum with sufficient accuracy to predict the winning ball, let alone all the molecules in the universe. So, for all intents and purposes, the future is unpredictable.

This explanation did not satisfy many who objected that Free Will 'for all intents and purposes' was not Free Will at all – merely the illusion of Free Will.

Then along came quantum theory and with it, Heisenberg's Uncertainty Principle. (See the section of the same name on page 169) This seemed to suggest that there was an inherent randomness in nature which was easily sufficient to provide the tiny discrepancy in initial conditions which would make the future of the universe genuinely unpredictable. The problem is, if the future is random, how can a conscious mind with Free Will direct its course?

We seem to be between a rock and a hard place. Either the future is predetermined or it is completely random. In neither universe is there any room for Free Will.

Randomness

Many computer programs which attempt to look intelligent use a random number generator to make it appear that they are making choices. ELIZA was a famous early example of a program which simulated the responses of a psychotherapist asking a patient questions. The program would extract certain key words from the patients response and, using a random number generator, select a new question, inserting the key word in an appropriate place. A typical exchange might go like this:

Patient: Well, my boyfriend made me come here.

ELIZA: Your boyfriend made you come here?

Patient: He says I'm depressed much of the time.

ELIZA: I am sorry to hear you are depressed.

But if the computer had generated some different random numbers the exchange could equally well have gone like this:

Patient: Well, my boyfriend made me come here.

ELIZA: Why did your boyfriend make you come here?

Patient: He says I'm depressed much of the time.

ELIZA: Do you think your depression has anything to do with your boyfriend?

Now most computers use a very simple method of generating random numbers. In fact they are not random at all but are generated by a mathematical process such as the following: start with a 10 digit number called the 'seed'; square it to obtain a 20 digit number; chop off the first and last 5 digits to obtain a second 10-digit number; repeat this process over and over again.

This algorithm has several problems. How do you choose the 'seed'? How do you know that the process won't enter a loop, generating the same sequence of numbers over and over again? And, of course, if you happen to start with the same seed that you used once before, you will generate exactly the same sequence of 'random' numbers. This is why such sequences are called pseudo-random.

If you really want to generate truly random numbers (and there are good reasons why statisticians and other professionals need them) you have to resort to monitoring truly random physical events such as the electrical noise generated by a warm diode (thermal randomness) or, better still, the radioactive decay of an element (quantum randomness).

Now it is possible – some would say likely – that the human brain is just a sophisticated computer with a random number generator inside which, like ELIZA, gives the appearance of making decisions; and either the random generator is algorithmic (in which case, all our actions are predetermined from the moment we are born) or physical (in which case we still have no actual control over our future actions). Either way, there is no room at all for anything which I would call Free Will.

I would like to offer an alternative suggestion.

Structural Dualism

A glance the figure on page 59 will reveal another 'ism pinned to the column which I have called Structural Dualism. This is the belief that the brains of human beings (and other conscious creatures) possess unique structures which carry out processes as yet not fully understood by science – the implication being that it is these processes which are responsible for our conscious awareness and also of our belief in our ability to determine the future to some degree.

From what I have said in the previous section, no scientific theory, whether classical or non-classical will, in my opinion. ever explain the subjective *experience* of consciousness; but some interpretations of Quantum Theory do open the door to at least the possibility of explaining how a conscious brain can exercise *free will*. It might work like this:

It is now well established that small systems consisting of one or two particles or even much larger systems at very low temperature can exist in several different states at the same time. (See **Superposition** on page 143). Opinions differ as to how large a system can be put into an entangled state and at what stage, if any, such a state reverts back to being normal (See **The collapse of the wave-function** on page 153) but I hope and believe that eventually a physical process will be discovered that explains how and why this collapse occurs. (See **Chaotic collapse** on page 177). I also believe that, when this process is discovered, it will show us how really large objects such as a human brain can enter an entangled state in which the whole structure is employed in some sort of quantum calculation. If the outcome of that calculation includes the possibility that a cup of tea is made, and also the possibility that a cup of coffee is made, and if the quantum processes which are going on in my brain are at the same time somehow responsible for my feelings that '*I*' am making a conscious '*decision*' then surely we are justified in saying that when the wavefunction collapses and I end up making a cup of tea, the '*I*' in my brain has exercised its '*free will*'.

The usual objection to an argument of this sort is that, if the decision is made by any sort of natural, physical process, whether it is quantum or classical, the result must still be either determined in advance or completely random. In neither case is it possible to claim that my brain has exercised its free will in any meaningful sense.

My answer to this is as follows: Nobody could predict the outcome in advance *even in principle* (because the process is a non-classical one) so the process is not deterministic; but neither is it random because the whole, conscious brain (which is the 'I' bit) was definitely responsible for causing the one outcome and not the other. The situation can be likened to a General Election. The outcome is unpredictable, but it is not random either because the whole electorate is involved in making the decision.

Some will object that I am just mincing words here but I do not believe that that is so. If conscious creatures possess this thing called free will and unconscious creatures do not, then it ought to be possible to detect this difference in their respective behaviour. To some extent we have already discussed this with regard to the evolutionary benefits which consciousness confers but is there any way we can detect whether or not a creature has *free will*? With humans, it is not a problem. When my wife says 'I am going to make a cup of tea.' and then subsequently makes a cup of tea, we can be reasonably confident that she has made a conscious decision to carry out that action. (Of course, a computer-controlled robot could be programmed to do the same thing in in specific circumstances but in an open ended Turing-style test, its limitations would soon become apparent.)

With creatures that cannot speak, the task is more difficult but we are still looking for the same thing – evidence that the creature knows what it wants to do before it does it. The dog that drops a ball at its owner's feet clearly knows what it wants to do and his posture speaks of his intentions as loudly as words. The squirrel that looks around and notes the positions of the nearby trees as it hides a nut in the forest does so because it knows that it may need to return to the same place in the future. The Jay that attentively watches the squirrel hide the nut does so because it intends to dig up the nut as soon as the squirrel is gone. All these creatures appear to know what they are doing before they do it. They all show evidence of intent.

But does the spider know what it is about to do when it lays down the first strand of a web? I doubt it because there is no need for it to know. Web-building is an instinctive pre-programmed task. When a pike disguises itself among some rocks, is it thinking to itself 'I will just hide in here until a minnow comes along so I can eat it' I don't think so. Its behaviour is entirely instinctive. When a chameleon crawls onto a leaf and turns itself green, does it make a conscious decision to do so? I doubt it. None of these creatures ever show evidence of intent. Of course, it has to be admitted that none of these creatures have the means to show us that evidence so the argument is far from conclusive but I cannot believe that it is merely coincidence that the creatures listed above which do show evidence of intent are precisely those creatures which show REM / non-REM cycles of sleep. The reason must surely be that it is because their brains have the ability to use quantum (or other non-classical) processes to make free and creative decisions about what to do next. In other words, these creatures are the ones which have the capacity of conscious thought.

So to explain the existence of Free Will in human beings (and presumably in other conscious creatures too), we therefore require three things:

A) that quantum (or other non-classical) processes are important in the way neurons work;

B) that the firing of individual neurons can have macroscopic consequences and

C) that certain holistic patterns in the brain can in turn influence events on a quantum scale.

We can illustrate the interconnectedness of these ideas in the following way:

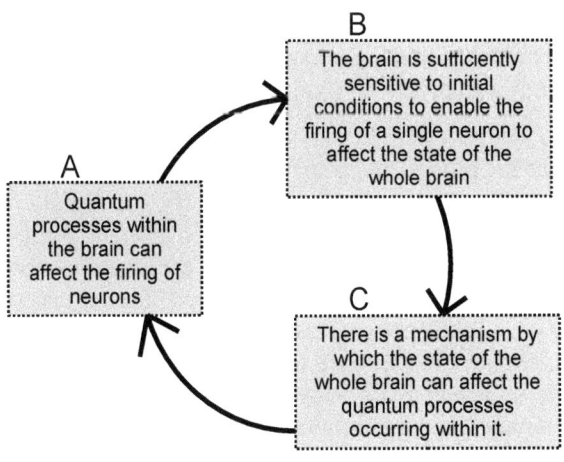

Proposition A – that quantum processes are important in the way neurons work – is highly controversial. Small though it is, a single neuron contains billions of atoms and although the behaviour of individual atoms in its synapses may involve quantum processes, most neurologists would say that they have no bearing on whether the neuron ultimately fires or not and the behaviour of a neuron can be completely described by a purely classical model. Roger Penrose has pointed to certain auxiliary structures in the brain called microtubules which may depend on quantum processes but it has to be said that his ideas have not gained universal acceptance.

Proposition B – that the firing of a single neuron can influence the state of the whole brain at a later stage – is much less controversial. We have all heard of chaos theory and are familiar with the idea that the flapping of a butterfly's wing in Australia could later cause a hurricane in Florida. (As a matter of fact, I am extremely doubtful about that particular claim as, chaotic though our weather systems might be, I doubt if it is quite that chaotic.) Nevertheless, the principle remains and few people doubt that an object with the complexity of the human brain is perfectly capable of behaving in an extremely chaotic way if it is wired up to do so.

The really controversial claim is proposition C – that macroscopic states of the brain can influence quantum events within individual neurons. Without this claim, the workings of the brain are entirely bottom-up and it is seen to be a mere calculating machine. Fortunately, the idea of suspended reality and chaotic collapse provides us with a mechanism whereby the loop can be completed, and conscious brain states can influence the subsequent behaviour of the whole organism.

I imagine the process to go something like this. It is 11 o'clock and my brain starts to receive signals from my body telling it that blood sugar levels are running low. This sets in motion a series of habitual (hard-wired) sub-processes which cause me to get up and go to the kitchen cabinet where the tea and coffee is stored. The regular sub-procedures take me to the point where my hand is poised to pick up one of other of the two jars. Now I have to make a conscious decision. My non-classical brain enters into a holistic calculation in which all sorts of past experiences are taken into account – which one I chose yesterday, the cost of coffee, whether there will be enough left for tomorrows coffee morning etc. etc. For a split second reality is suspended while the wavefunction computes, then chaotic collapse occurs and the decision is made: I will have tea today and signals are passed to my arm accordingly.

Free Will - A serious objection

But for many die-hard materialists, this idea doesn't really change anything. Even if we accept propositions A to C, the old argument about the laws of Physics being either deterministic or random still applies. Free Will is still an illusion. The argument goes like this:

- All the propositions above come about through the action of some known *or unknown* law of Physics.

- By definition, a law of Physics must either be deterministic or at least, probabilistic – that is to say, the outcome of a physical process must either be fixed or one of a range of possible outcomes whose total probability is one.

- Since the human brain is subject to the laws of Physics, its behaviour too must be either deterministic or random.

- Neither deterministic nor random behaviour is compatible with the idea of Free Will.

What are we to make of this? It looks foolproof. All this talk of suspended reality and non-classical laws of Physics which govern how brain states can affect individual neurons counts for nothing because, at the end of the day, a brain is still just a calculating machine whose behaviour is partly pre-determined and partly random. It makes no difference at all whether the system passes through a conscious state or not; its apparent decision to move its hand one way or the other is either pre-determined or random. Brains, even conscious ones, cannot possess Free Will. Free Will is an illusion.

It may surprise you to learn that, at the end of the day, I am prepared to accept this argument, at least in principle. What I object to is the use of the phrase 'just a calculating machine'. The conscious brain is far from being 'just a calculating machine' mindlessly obeying the laws of Physics. Yes, it obeys the laws of Physics – but not 'mindlessly'. That is the whole point. The conscious brain has to be classed as a *mind* and exists in a different metaphysical world from the world of 'mere calculating machines'. Consider the following analogy.

When the people of Scotland voted in 2014 to reject independence, you could argue that the outcome was either predetermined (by all the voters prejudices) or random (because most of the voters hadn't sufficiently considered all the factors involved in the issue). But to do so would be to deny the existence of an entity called 'The People of Scotland'. If you believe in democracy, them you must also believe that the process of discussion and debate which precedes an important vote produces a result which in some important way transcends the prejudices and foibles of the individuals who vote. In the same way, because the whole brain is involved in making *conscious* decisions, the outcome transcends that of a 'mere calculating machine'.

Nor is it simply a question of the brain being extremely complex. Epiphenomenalists such as Douglas Hofstadter would have us believe that the brain has so many feedback loops in it (what he calls a *tangled hierarchy*) that we must expect behaviour to emerge from it which cannot be deduced from a consideration of the Physical laws which its components obey. This is an important idea and worth considering in more detail.

It is true that feedback loops can produce new and sometimes unexpected behaviour from quite simple systems. Everybody is familiar with the characteristic howl that emerges from a loudspeaker when the amplification of a PA system is turned up too much and this is the sort of thing you can get if you point a video camera at a TV screen:

The laws which govern the frequency of the howls from the loudspeaker and the development of the swirling patterns on the TV screen are excellent examples of what are called *emergent phenomena*. (See: ***Emergent and Transcendent Properties*** on page 190 for more detail) They appear to have a life of their own but in fact, they can, in principle if not always in practice, be deduced from more fundamental principles. Now Hofstadter's view is that these feedback loops can become so complex that the resulting system can exhibit behaviour which cannot be deduced, even in principle, from more fundamental laws. In other words, in my terminology, a tangled hierarchy of feedback loops can be so complex that it can produce *transcendent* as opposed to merely *emergent* behaviour.

While this is a superficially attractive idea, I cannot ascribe to it. At what point, exactly, does a tangled hierarchy become sufficiently complex as to transcend the laws of Physics? How could a tangled hierarchy ever become conscious or exhibit Free Will? In my view, however complicated the feedback loops are, the behaviour which results is always going to be explicable in more fundamental terms at least in principle.

So how can we explain genuinely transcendent phenomena like consciousness and Free Will can exist in a universe which is governed by laws which are in part deterministic and in part random? This is the heart of the problem and I shall try to spell out my answer as clearly as I can once more.

My contention is that:

• Brains become conscious when they use some as yet unknown physical process.

• Even if we came to understand this physical process in detail, we would still, in all likelihood, be unable to explain how the process brings about the *experience* of consciousness.

• Notwithstanding this, I believe that when my brain makes a conscious decision to do something, the *whole of the conscious brain is actively employed* in bringing about the necessary changes in the patterns of neuronal activity which ultimately lead to the decision being carried out.

• It is possible that the way this is done is through the collapse of the wave function by an exquisitely arranged environmental configuration but even if this is completely wrong, we should at least apply the lessons we have learned from quantum theory as a guide.

• Following this lead, we can say that when a conscious decision to do something is made, the outcome is not determined in advance but neither is it random, because *the whole conscious brain was involved in the decision.*

To illustrate this theory with another example, suppose I feel cold and wonder whether or not to turn on the central heating. This is the sort of decision that only a conscious brain could make because it involves a huge amount of stored previous experience involving not only the mechanics of thermostats and gas boilers but also the state of my finances and the potential reaction of my thermo-phobic wife etc. etc. At some critical stage the whole brain becomes involved in some non-classical computation in which all these factors are weighed up. Two outcomes are possible at this stage – either I get up and throw the switch or I remain seated. By some mechanism as yet unknown, the wavefunction collapses and the decision is made: I get up and switch on the heating. The outcome was not determined in advance (because the processes involved are non-classical) but neither was it random (because I, that is – my brain, made the decision).

Now contrast this process with the decision made by the thermostat in the system a few seconds later. The thermostat feels cold. This sets up strains in the bimetallic strip inside it which cause the strip to flip and switch on the boiler. The process is, of course, completely deterministic because all the processes involved are classical. Even if we were to construct a quantum thermostat which, for example, switched on the boiler when a radioactive atom chose to emit an alpha particle, it would be ludicrous to ascribe free will to this device because the outcome is now completely random. Just employing non-classical processes does not make a device conscious.

It is only when the device is arranged in such a way that consciousness emerges from the non-classical processes that free will becomes possible. That is why it is only conscious animals which show evidence of intent, enjoy long-term personal relationships, design tools, tell jokes, compose symphonies and make mistakes. It is free will that sets humans – and many species of mammals and birds – apart.

Conscious machines

Nothing I have said so far rules out the possibility that we might one day build a conscious machine but in order to do this we will have to develop a theory of wave-function collapse and we will have to use this theory to construct a machine capable of maintaining itself in an entangled state for long enough to enable large parts of the machine to contribute simultaneously towards the eventual outcome.

I would expect such a machine also to possess free will – to exhibit evidence of intent, to possess long-term memories and be able to recognize itself as an individual. It might even be able to tell jokes and compose symphonies. But the big question is – would it have a 'self'?

The mystery of selfhood

If you can bring yourself to accept the hypothesis that all brains are basically hard-wired robots but that the brains of some higher-order animals can work in a second non-classical mode which gives rise to an transcendent property which we call consciousness, then pretty well all the paradoxes which have puzzled philosophers and lay persons down the ages simply evaporate.

We can see how it is possible for the same brain to be in different states of consciousness at different times and how different brains can have different degrees of consciousness. We can be absolutely certain that computers and many primitive brains are not conscious and we have, in principle at any rate, a foolproof test of whether a particular brain is capable of conscious thought or not. This should eventually be able to give us scientifically reliable answers to the questions of which other species of animals are capable of conscious thought and, most importantly, at what stage in the development of the human embryo this capacity emerges. We can see how the facility of conscious thought gives those higher-order animals an evolutionary advantage over their competitors and, from the fact that conscious thought seems to have evolved independently in two separate classes of animals, we can be confident that consciousness is not a miracle or a freak occurrence, but a natural consequence of the laws of physics (some of which are as yet unknown). We can even envisage a time when, those laws being made clear to us, we could (if we were foolish enough) build a conscious being out of silicon or some other inorganic material which would be able to compose symphonies, tell jokes, and even (horrible thought!) exercise its own free will to make war on mankind if it so chose to do so.

So have I answered all the questions which I posed at the start of this essay? Not quite. There is one outstanding issue left which, ultimately goes to the heart of the paradox of consciousness and that is – how do we actually *know* that we are conscious? The problem is this: we only know we are conscious *because we are conscious*. The circularity inherent in this argument is an indication that consciousness itself (as opposed to a non-classical process going on inside a brain) is not amenable to objective scientific analysis. Consciousness is a meta-phenomenon, outside the realm of scientific enquiry. Even if and when we come to understand the non-classical process that give rise to consciousness, we will still not *understand* consciousness in any meaningfully objective way.

So perhaps we should simply ditch the notion of consciousness altogether. Once we understand the workings of the human brain and can construct non-classical robots to compose symphonies for us, perhaps we should simply admit that that is all there is to it and that our subjective experiences of our own conscious thoughts are at best irrelevant, at worst an illusion. Perhaps we are all, if not mind*less*, then merely mind*ful* robots going about our business in accordance with the non-classical laws of physics, exercising our free will, loving and hating, telling jokes and making war, thinking irrelevant thoughts and wondering why we bother.

For many, such a soulless scenario is an abomination and I can see why many people will want to reject it out of hand. I, too, am uncomfortable with it but for a different reason. To see why, let us go right back to first principles.

In 1637, René Descartes published his Discourse on Method which contained the famous argument '*Je pense, donc je suis.*' or in English '*I think therefore I am.*' (It was only later translated into the more famous Latin '*Cogito ergo sum*'.) Descartes argued very persuasively that he could doubt absolutely everything else, but the one thing that he could not doubt was his own existence – because he knew he was thinking. In other words, because he was conscious. This argument has become so familiar that it is accepted almost without question and it can be said to be the foundation stone of every Western philosophy ever since. Indeed, if you just sit quietly and contemplate your own thoughts, you will quickly convince yourself that it is true. If '*I*' am thinking then surely '*I*' must exist – otherwise what or who is doing the thinking?

But with issues as important as this, we must tread very carefully indeed. I am reminded of the story about the grandfather, the father and the son who went for a trip in a train. The son was looking out of the window and suddenly exclaimed "Look, Daddy. All the sheep in that field are black!". The father, who was a careful man said: "Well you can't be sure of that. The best you can say is that all the sheep in the field *which you can see* are black." The grandfather (who was a mathematician) corrected him. "Actually the best you can say is that all the sheep in the field which you can see are black *on at least one side.*" Let us apply the same rigour to Descartes' famous phrase.

The first criticism we can make is that as soon as we have uttered the first word (Je or I) we have begged the question. This defect is easily remedied as follows. 'Thoughts exist, therefore there must exist an '*I*' which is having those thoughts.' For the sake of clarity, let us define a 'self' as that entity which has thoughts. We now have: 'Thoughts exist therefore selves exist.' I hope we can all accept the argument so far. '*I*' have thoughts so I am a 'self'. Presumably you also have thoughts so presumably you too are a 'self'. Since we believe that there are many billions of conscious creatures on this planet all having different thoughts, there must be many billions of different selves.

So far so good. But it is here that we begin to make unwarranted assumptions. When I go to sleep, I lose consciousness. And when I wake up the next morning I regain consciousness and take up my life apparently where I left it the night before. It is natural to assume that the 'self' which wakes up in the morning is the same 'self' that went to sleep the night before. But this does not follow from Descartes' aphorism. To be strictly accurate, the best we can say is "Thoughts exist therefore selves exist *while the thoughts exist*." We cannot deduce that selves have an independent existence separate from the thoughts which the selves have.

'Oh but this is ridiculous!' I hear you say. 'When I go to sleep, it is still *me* who wakes up in the morning – not some other self! That would be absurd!'

With the greatest respect, it would not be absurd at all. If a 'different self' were to occupy your body when it wakes up in the morning, it would wake up remembering that it had enjoyed last night's concert (in which you took part) and that it had a dentist appointment at 2 o'clock (which you were not looking forward to) and that Auntie Mabel was coming to tea (which you were dreading) etc. etc. In short – it would still be 'you'. Similarly, if 'your self' woke up inside someone else's body, it would not have any recollection of it being 'you' the day before. It would simply *be* someone else.

In fact, it is the assumption that selves have an independent existence separate from the body it occupies which lands us in so much trouble. Consider the Spock paradox.

When Scottie beams Dr Spock up to the starship Enterprise, all his molecules are scrambled, sent up in a plasma beam and reconstituted inside the spaceship. How can we be sure that the beamed-up Spock is really the same as the original? How does the 'self' get transported? When exactly does the 'self' leave the original body and enter the new one?

Worse still, consider the Kryptonite Man paradox. Kryptonite Man is an exact clone of Superman created by making an exact copy of Superman molecule by molecule. Does Kryptonite Man share the same 'self' as Superman or are there two 'selves'? Suppose we create two cloned Supermen, destroying the original in the process. Which clone would claim to be the 'real' Superman?

And if you think that these examples are too contrived and only prove that teleportation and macrocloning are not possible even in principle, then what about the experiences of patients with schizophrenia or those whose brains have been surgically divided into two? What happens to the 'self' of someone who has Alzheimer's disease or who suffers an accident which completely changes their personality? Is it possible for a (non-physical) self to change as a result of a physical accident? How many different selves can inhabit the same body?

As soon as we abandon the idea of a 'self' having an independent existence, all these questions are answered. Every time the non-classical processes inside a conscious brain are switched on, a 'self' comes into being – and when the conscious brain goes to sleep or dies, the 'self' disappears. There is no contradiction in the idea that a patient with a 'split brain' could think two thoughts at once, or that one half of his brain is unaware of what the other half is experiencing. Nor do we need to be surprised that personalities can change over time, sometimes depressingly so. If the chemistry of the brain changes, personalities may change also. Ultimately, we can see that the concept of a 'self' is a completely empty one and that Descartes' sentence is actually nothing more than a definition and no more or less true or meaningful than the statement 'Mistakes exist therefore boojums exist.' (where a boojum is defined as a thing which makes mistakes!)

But, dear reader, I sense that you are still not satisfied. What is it that provides the essential sense of *continuity* which we all experience when we wake up in the morning, the sense that I am still the same *person* who went to sleep last night. What is it that gives me the unshakable feeling that inside this mortal body there is a real *me* which inhabits it? Why do I feel as if I inhabit *this* body as opposed to someone else's? Believe me – I am troubled by these questions in exactly the same way as you are. I am not going to write it all off, as many materialist philosophers have done and say that consciousness in nothing but an illusion. But neither do I believe that there is a *me* which is in some way independent of the body which is typing this essay. The thing which actually defines *me* is nothing more than my body, my brain, and all the neuronal interconnections in it which have been forged over the last 60 years and which constitute my memories and my personality. It is this mortal body which provides the continuity between one day and the next and that is all. If you cannot rid yourself of the notion that there must be a 'self' which is thinking today's thoughts, that is OK because that is the way a 'self' is defined. But whether today's 'self' is the same as yesterdays is a meaningless question. Is the telephone number that you use to phone your mother today the same as the number you used yesterday? I don't mean, is it the same number but is it *the same actual number*? The question is pointless. Mathematically the two numbers are identical but it makes no sense to argue that the numbers are the same in any physical sense.

In truth, it is not a 'self' which thinks thoughts, it is a conscious brain operating in non-classical mode. And the brain that wakes up today is, of course, the same as the brain that went to sleep yesterday.

But I am still hearing howls of protest from my exasperated reader: "OK so my brain provides the continuity from one day to the next but you still haven't explained why I feel that I am so *unique*. Why am *I* here, now, reading this and not someone else, over there playing football? When I look round at all the other people around me I can accept that they are all *like* me but they are not ME! *I* am ME and no one else is!"

Yes, I do see your point. But then I feel exactly the same as you do. So does every other conscious being on the planet. But then that is really my point also. We are all equally unique. The only way to avoid this kind of pointless argument is to reject the idea of a 'self' altogether and restate the case in purely physical terms.

Consciousness, I have argued, is a meta-phenomenon which arises naturally when brains (and/or possibly other structures such as quantum computers) use non-classical processes to make free decisions about what to do on the basis of long-term memories of their past experiences. Since we do not yet know enough about the non-classical processes referred to, we are, as yet, unable to construct conscious machines and even if we reach that state of enlightenment, we will still not *understand* why or how the system we have built is conscious. What we do know, however, is that whenever a system is conscious, it has a powerful sense of being a unique individual. This is not an accident or an unnecessary by-product, it is an essential feature of consciousness and it is the reason why conscious individuals can relate to other conscious individuals in a unique way. This ability to recognise oneself and others as individuals is what enables penguins to find their mates after months of separation at sea; chimps to show their offspring how to make tools to raid an ants nest; elephants to mourn over a lost parent and humans to fall in love.

Every time you wake up in the morning, you do so with this overwhelming sense that your are unique and the same person that went to sleep. You are not wrong. But it is not some disembodied 'self' which is unique and which endures from one day to the next – it is your body and your brain which has been endowed with this miraculous capacity of conscious thought which endures through the night and wakes the next day.

I said earlier that I was uncomfortable with the idea of

consciousness being merely an illusion and that we humans are still robots after all – just robots with attitude. The reason for my discomfiture is not that I dislike the idea that we are 'robots with attitude' because that us exactly what we are; fabulously complex machines with the most amazing power to understand the world around us and our own unique place in it and to direct its future. What I object to is the implication behind the use of the words 'merely' and 'illusion' as if there is something to be ashamed of in saying that we are 'robots', or that our 'attitude' is some how not real and illusory. The faculty of consciousness is the most glorious, wonderful, colourful, exciting, exasperating, uplifting, extraordinary, emotional, inspirational, incomprehensible illusion anyone could possibly want. Why should we ask for more?

Conclusions

The theory of Mind developed above involves, not just a specific interpretation of Quantum Theory, but a new mathematical model which provides a mechanism for chaotic collapse and also the possibility of constructive interference from structures far from the place where the quantum event is taking place. Is there any realistic prospect of developing such a model?

I am afraid the answer to that question is probably No. In a sense, that is not an issue because it will probably turn out to be just as difficult to prove that such a model cannot exist as to prove that it does. For me, the logic which has led me to this conclusion is sufficient to persuade me that such a model does exist even if we cannot possibly figure it out; for others, the apparent absurdity of patterns in the brain affecting matter at a distance through quantum events will cause them to reject the concepts of mind and Free Will as illusions; and, of course, there will be many who don't see what all the fuss is about and just get on with life regardless.

We should also ask ourselves whether the theory has any practical, even testable consequences?

Again, I suspect that the answer is no. I do not see any realistic prospect of building a conscious machine as described earlier. Nor do I see any realistic prospect of us being able to study the human brain in sufficient detail as to identify any of the proposed patterns which constitute conscious thoughts. The numbers of neurons which are involved in the most trivial of brain activities is mind-bogglingly large. Indeed, there are probably no trivial brain activities. As far as the brain is concerned, we are like the alchemists who were fascinated by the behaviour of substances when they were heated or mixed together but had no way of visualising the atoms and molecules which made sense of all these effects. A statement like 'brains become conscious when they use quantum effects to process information' is really no better than 'metals give off phlogiston when they are heated to a sufficient extent'. Both statements in their day are capable of providing equal mental satisfaction, but neither can be said to provide the whole answer.

Notwithstanding what I have said above, I believe that the framework I have outlined seems to me to be logically consistent, to provide an intellectually satisfying answer to the mystery of consciousness and to provide a workable basis for an ontology which has no need for a God or a human soul but which nevertheless ascribes to human beings a unique ability which is shared by only a relatively small number of other species, namely the power to exercise our Free Will. And it is in the sagacious exercise of this power that the destiny of the human race will lie.

Fourth walk

Alan and I were strolling through the Royal Game Park on the island of Djurgården watching the ferries threading their way through the maze of islands that separates Stockholm from the open sea.

"Do you remember that fabulous day on Catbells?" Alan remarked. "That was when it all began, wasn't it?"

"Yes, I suppose it was. We spent the whole day talking about space and time and things. Do you remember the Tornado that shot past us up Borrowdale?"

"Yes I do. That was when I first realized that Relativity was all just about points of view. And then there was that incident on Scafell Pike. I couldn't help laughing! You were so goddamn *angry*!"

"Well, I had every right to be. You didn't seem to be a bit concerned about me. You only seemed interested in lunch."

"But I gave you the idea, didn't I"

"Yes. You did."

"That was when you thought up the idea of suspended reality and chaotic collapse wasn't it?"

"They don't call it that now. Its called 'Holistic Decoherence Theory' or HDT"

"Well you wouldn't have thought it up if you hadn't got lost in the fog."

"No thanks to you, there, though."

"But it was me who gave you the idea of linking holistic decoherence with consciousness too, while I were in hospital."

"Yes, you did that as well."

"So I should have received the Nobel Prize – not you!" He said.

"Well, you certainly deserve some credit. And don't forget, I shared the prize with that brilliant Polish mathematician who worked out all the maths and the team of quantum scientists in California who proved that the quantum Zeno effect really did occur, just as my theory predicted."

"What are you going to do with the money?"

"It is going towards the founding of a new institute. They are going to call it LIRAC – the Linton Institute for Research into Artificial Consciousness"

"Impressive."

"You know, they have already found evidence that holistic decoherence plays an important role in the workings of an ant's brain and there is more than a suggestion that the whole ant colony could be involved in the decision making process in, for example, moving the queen."

Gradually the Scandinavian sun slid towards the horizon, bathing the islands and the golden city in its gentle radiance.

Ah well. Everyone can dream . . .

Exploring Further

The four dimensions of Space-time

I have said that space-time has four principle dimensions and that the temporal dimension is different from the other three. It is a mistake to think that Special Relativity makes no distinction between space and time. On the contrary, it make the differences absolutely explicit. Let me explain.

Two observers in relative motion view space-time from different angles and therefore have a different perspective on events just as two people sitting at opposite ends of a table will draw a different picture of the table before them. But just as our two diners will get the same answer when they calculate the distance (by using laser ranging and Pythagoras' theorem!) between the salt and pepper pots, so our two relativistic observers will calculate exactly the same 'interval' between two points in space-time. Now in special relativity, the 'interval' (I) between two points is also calculated using Pythagoras' theorem. but instead of using the formula

$$R = \sqrt{(x^2 + y^2 + z^2)}$$

In space-time, the equivalent formula (in units where the speed of light is 1) is:

$$I = \sqrt{(x^2 + y^2 + z^2 - t^2)} \quad {}^8$$

and the difference between space and time lies in that crucial minus sign. We could write the formula equally well as:

$$I = \sqrt{(x^2 + y^2 + z^2 + (it)^2)}$$

where the temporal dimension is written as an imaginary number (i being the square root of minus one).

So time is indeed a fourth dimension, but it doesn't just lie at right angles to the other three in a kind of four dimensional space – it lies in the direction of imaginary space and the mathematics of time is crucially different from the mathematics of space.

---

8   For technical reasons I is usually defined as $\quad I = \sqrt{(t^2 - x^2 - y^2 - z^2)}$
    but this does not make any difference to the argument

Past and Future in Special Relativity

The idea that spaceships get shorter, clocks go slower and electrons get heavier the faster they move is now so well-known that it is almost possible to say that these, once bizarre effects, are now accepted by the general public, even if they are not understood.

The idea, however, that moving in a fast spaceship could reverse the order in which two events happen would still appear incredible to the great majority of lay people. After all, if a lightning strike causes a tree to fall, you can't make the tree fall before the stroke of lightning just by getting into a fast spaceship.

That is true, well enough, but let us consider the case of an astronomer who happens to be observing Jupiter through a telescope during the thunderstorm. He sees and hears the flash of lightning outside his observatory exactly on the stroke of midnight and then, a few minutes later, sees a stroke of lightning on the surface of the giant planet. Knowing, as he does, that Jupiter is at the time about 10 light-minutes away (i.e. about 112 million miles) he reasons that, since it took the light 10 minutes to reach him, the stroke of lightning on Jupiter actually happened a couple of minutes *before* the lightning stroke on Earth.

Now just at the same stroke of midnight, an interplanetary space cruiser happened to pass by Earth travelling at high speed *away* from Jupiter. The captain of the cruiser noted the flash of lightning on Earth as he flashed by and also the time on his ships clock.

Now I do not wish to go into the mathematical details here, fascinating though they are. Suffice it to say that, owing to the way that the equations of special relativity work, when the light from the flash of lightning on Jupiter eventually catches up with the spaceship, the time on the ships clock reveals that the Jupiter flash occurred several minutes *after* the flash of lightning on Earth.

Now the only reason this can come about is because although the two lightning strikes are well separated both in space and time, they are sufficiently close together to make it impossible for light to travel from one to the other in the time available.

It is a lot easier to see this on a Space-time diagram:

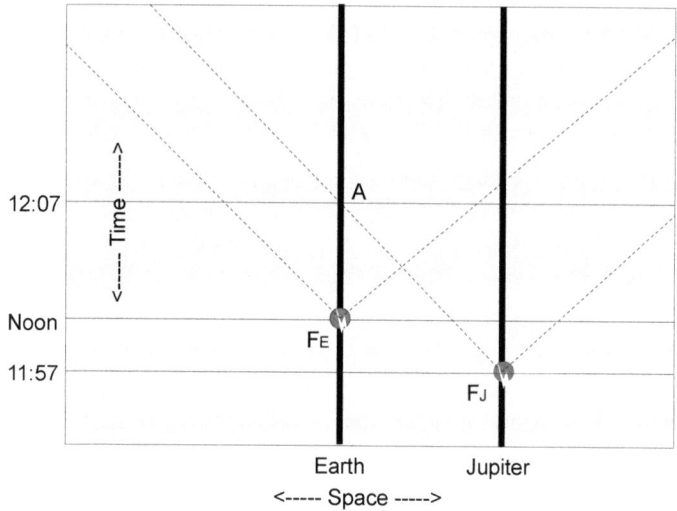

Here we have Space plotted on the X axis and Time on the Y axis. Earth and Jupiter are represented by two thick black 'world lines'. The two events we are interested in (the lightning flashes on Earth and Jupiter F_E and F_J respectively) are plotted as points on the diagram. The light itself from these flashes is plotted as 45° dotted lines and the horizontal lines represent 'now' as seen by the Earthbound astronomer. The arrival of the light at Earth from the flash on Jupiter is labelled 'A'. Naturally, although the event A occurs after the event F_E, the astronomer has no difficulty in inferring that F_J happened *before* F_E.

Now let us consider the situation from the point of view of the captain of the spaceship.

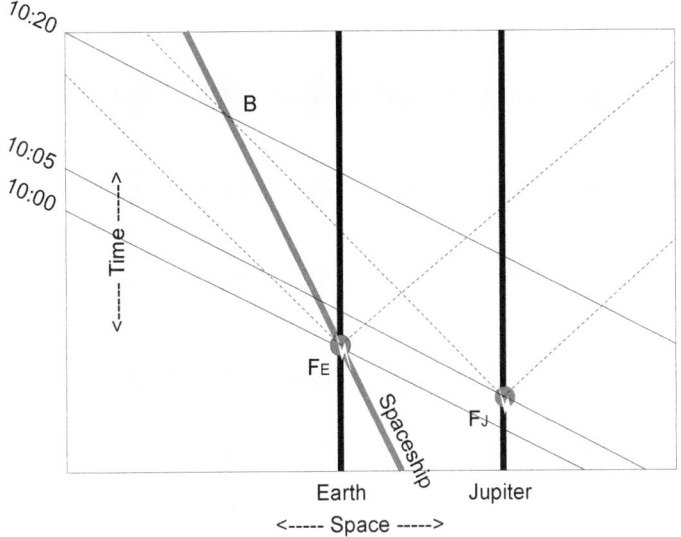

The spaceship is moving *away* from Jupiter so it is represented by the thick grey line slanting to the left. This line intersects the Earth line at the point F_E (Noon on Earth, 10:00 spaceship time).

Now here comes the tricky bit. Special Relativity demands that the line which represents 'NOW' must always make the same angle with the speed of light (the dotted line) as the world line of the observed. This means that the more the spaceships world line slants to the left, the more the spaceships 'NOW' line (the one labelled '10:00') slants upwards. When the captain of the spaceship observes the flash of light from Jupiter (i.e. at B, 10:20 by his clock), his calculations lead him to infer that F_J occurred at 10:05 i.e. *after* F_E.

Now a moments thought will lead you to the conclusion that, if you go fast enough (but always less than the speed of light of course), you can reverse to apparent order of any two events *provided that, like F_E and F_J, neither lie in the future or past light cones of the other.*

What this means is that every observer can put every event in the universe into one of 5 different categories. There are events which definitely happened in the past; events that will definitely happen in the future; events elsewhere in the universe which look to me (but not necessarily to anyone else) to be happening right now; events which I think haven't happened yet (but others might) and events which I think have already happened (but which others might think haven't happened yet).

Again, a diagram will help.

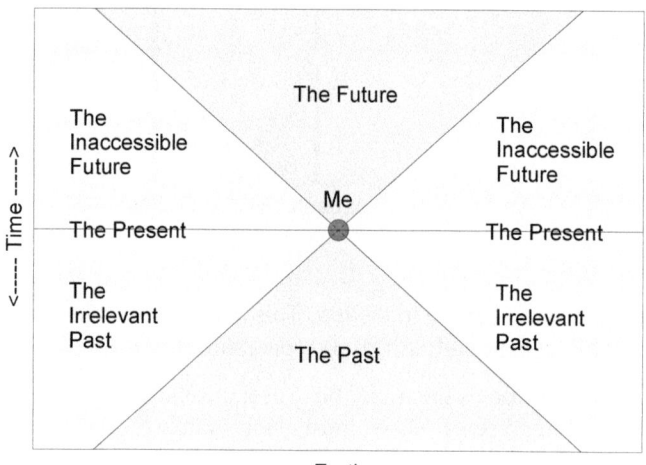

I call the region which as far as I am concerned hasn't happened yet the *Inaccessible Future* because nothing I can do can influence these events – because to reach them I would have to travel faster than light.

Likewise I call the group of events which I think have already happened the *Irrelevant Past* – because although these events have, in my opinion, already happened, they cannot possibly have had any effect on me, now, because any influence they might have can only travel at the speed of light.

Special Relativity and Causality

If you have read the previous section (Past and Future in Special Relativity) you will already be 90% towards seeing why, even though by travelling in a fast space ship you might be able to reverse the temporal order of some events, Special Relativity is perfectly consistent with the logical condition that if an event A *causes* and event B, then A must have occurred *before* B to every possible observer.

The only proviso that we have to add is the absolute prohibition of any *thing* (whether material or immaterial like a thought or a photon) travelling faster than light.

It is well known (and well verified experimentally even by the GPS device in your pocket) that when objects travel fast, time travels more slowly. When an astronaut travels in a fast spaceship to a distant star and back again, he returns significantly younger than his stay-at-home twin brother. Special Relativity predicts that if you travel at the speed of light, time will cease altogether and we may fancifully suppose that if you travel faster than light, time will travel backwards and you will return, not merely younger than your twin brother, but younger than you were when you left!

This is obviously nonsense.

Provided that material objects travel at less than the speed of light, and immaterial objects like photons and information travel at no more than the speed of light, causality will be preserved. (This is why there was so much scepticism about the recently announced discovery of neutrinos which apparently beat photons in a race from Cern in Switzerland to a laboratory in Italy.)

Apparently, there could exist particles which *always* travel faster than light but it would be impossible to use them to communicate information. As far as I know, nobody has seriously suggested that they *do* exist and there is certainly no experimental evidence for them.

Randomness, Order, Disorder and Entropy

Have a look at the following number (or sequence of digits):

```
1,415,926,535,897,932,384,626,433,832,795,02
8,841,971,693,993,751,058,209,749,445,923,078,16
4,062,862,089,986,280,348,253,421,170,679
```

Would you say this was a random number? It looks pretty random. It contains 8 zeros, 8 one's, 12 twos, 11 threes, 10 fours, 8 fives, 9 sixes, 8 sevens, 12 eights and 14 nines which is pretty much what you would expect from a random sequence of digits. But this sequence is far from random (as you may have noticed from the first few digits). It is, ion fact, the first 100 digits of the decimal expansion of π (after the decimal point).

So what is the difference between a truly random number and one that just *looks* as if it is random? The answer is this: if a number (or sequence of digits) can be specified by a description which is shorter than the number itself, then it contains some degree of order and cannot be described as truly random. For example the following number

```
1,313,936,333,837,333,334,323,433,333,735,32
3,831,373,633,393,731,353,239,343,435,323,038,36
3,032,363,039,383,230,343,233,323,130,373
```

can be described as 'the first 100 digits of pi with every other digit replaced by a 3'.

The next step is to try to quantify exactly how random a given number is. It was Boltzman who showed us how to do this. He defined the entropy of a system as the logarithm[9] of the number of possible states of the system which share the same macroscopic properties divided by the total number of possible states of the system. We can define the 'randomness' of a 100 digit number in a very similar way as the logarithm of the number of 100 digit numbers with a similar shortest description divided by the total number of 100 digit numbers.

Applying this rule, the first number is pretty well unique and

---

9   More correctly, the entropy of a system is defined as $k \ln(\Omega)$ where k is a constant called Boltzmann's constant, Ω is the ratio of the number of states as defined above and ln is the *natural* logarithm rather than logarithm to base 10. For simplicity we shall here ignore k and use the ordinary logarithm to base 10.

therefore has a randomness of $\log(1 / 10^{100})$ which turns out to be -100.

As for the second number, it shares the same general description with at least nine other numbers, one of which is 'the first 100 digits of pi with every other digit replaced by a 7'. Its randomness is therefore equal to $\log(10 / 10^{100})$ which is -99.

What about the following number:

1,415,926,535,390,561,734,.......

which has the following shortest description: 'the first ten digits of the decimal expansion of pi followed by three, nine, zero, five, six, one, seven three, four etc. etc. etc.' (the last 90 digits are truly random and have no pattern to them at all and so have to be spelled out individually)

In this case there are approximately 10^{90} other numbers which have the same general pattern to their shortest descriptions and so the 'randomness' of these numbers will be $\log(10^{90} / 10^{100})$ which is -10.

In the limit, the randomness of a completely random 100 digit number will be $\log(10^{100} / 10^{100})$ which equals 0.

Now it has to be admitted that this is a very peculiar scale. Our unique number has a randomness of -100 while a perfectly random number has a randomness of 0. You could, if you like, ignore the minus sign and call the quantity 'order' rather than 'randomness'. Alternatively, you could arbitrarily add 100 and get a scale which starts at 0 and goes up to 100; it doesn't really matter. What is important is that there exists a reasonably objective measure of the randomness of a number.

In exactly the same way it is possible to quantify the amount of order or disorder in any physical system such as a container full of gas by counting the number of states which share the same macroscopic properties and taking the logarithm. This quantity is called *entropy*.

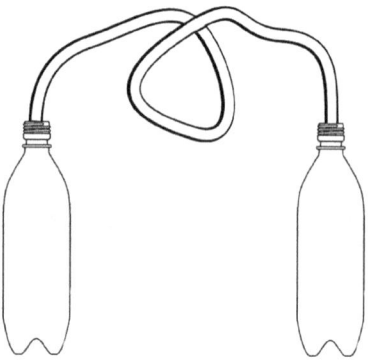

For example, suppose you have two sealed bottles connected together with a pipe containing, say, 1000 molecules. If the molecules are randomly distributed between the two bottles, then the probability that any particular molecule is in bottle A will be 0.5. The probability that *all* of the molecules are in bottle A will be 0.5^{1000} which is roughly equal to 10^{-300}. What this means is that, of all the 10^{300} ways there are of distributing 1000 molecules between the two bottles, only 1 has all the molecules in one bottle. The entropy of this system is therefore equal to $\log(10^{-300})$ which is -300.

If you start with all 1000 molecules in one bottle and then open a valve in the connecting pipe, molecules will spontaneously rush into the empty bottle because of their random thermal motion and eventually both bottles will contain roughly equal numbers of molecules. (For more on this point see: *Random Walks in Phase Space* on page 122.) Now the number of ways in which 1000 molecules can be divided into approximately two equal groups is staggeringly large. In fact it is not a lot smaller than the total number of ways in which the molecules can be divided into two groups. What this means is that the entropy of the system jumps up from -300 to nearly 0 in an instant. To see where this leads us, have a look at *The Second Law of Thermodynamics and the Arrow of Time* on page 114.

It is even possible, in principle, to calculate the entropy of the universe. Sir Roger Penrose has estimated that the total number of possible states in which the universe could have started is of the order of 10 to the power of 10^{123}. (If you thought that 10^{123} (i.e. 1 with 123 noughts after it) was a large number, this is just a tad larger. Written out, it would start with a 1 followed by 10^{123} noughts. Since there are thought to be only about 10^{80} atoms in the universe, you couldn't even write it out with one digit on each atom!) If we suppose, as seems likely, that the universe started out in a unique configuration with maximum order, this implies that the entropy of the universe when it started was -10^{123} – a stupendously low value from which it has been crawling its way upwards ever since.

The Second Law of Thermodynamics and the Arrow of Time

If, as I have said, our conscious minds dictate that time exists and that it flows inexorably in one direction, why is it that all the physical laws which we have discovered so far are time-symmetric?

Well, it is not quite true to say that all physical laws are obviously time-symmetric. The laws of mechanics are. If you take a film of two billiard balls bouncing off each other and run the film backwards, the reversed collision looks perfectly plausible; if you reversed a film of the planets orbiting the Sun, it, too, would look perfectly natural; so, too, would a film of some water waves in a bath. At least, the films would look plausible as long as they were not too long. Eventually, friction will cause the billiard balls to come to rest, the collisions between planets and comets will eventually cause the planets to spiral into the Sun and water waves will eventually decrease in amplitude. In a feature-length reversed film, billiard balls would spring to life, planets would emerge from the Sun fully formed and water waves would start at the edge of a pond and converge towards the middle. These observations lead to the formulation of one of the most famous and most misunderstood laws of physics – the Second Law of Thermodynamics. In one form it can be stated in the following way: in any sufficiently large closed system, entropy must always increase with time.

Two things must be stated straight away. The second law is completely different from the First Law of Thermodynamics which states something apparently very similar, namely that in any closed system the total energy remains constant. If we are being precise, the Second Law should say something like: in any closed system, the probability of entropy decreasing becomes vanishingly small as the number of degrees of freedom in the system increases. The Second Law is therefore a statistical law, not a basic principle of nature.

Secondly, unlike energy, the concept of entropy is very difficult to define. It is often described as the amount of disorder in a system – but 'disorder' is a very relative concept. What looks like disorder to a parent looks like perfect harmony to a teenager. More importantly, while a single molecule can possess a well defined amount of energy, it makes no sense to talk about the entropy of a single molecule. In fact, entropy, like other macroscopic quantities like temperature and pressure, only has meaning when applied to large numbers of entities.

Notwithstanding these two comments, how do we reconcile the existence of a temporally dependent law with the fundamentally time-symmetric physical laws from which it appears to spring? Ludwig Boltzman, one of the greatest mathematicians of the nineteenth century, battled with this paradox and was devastated when his supposed proof that entropy always increases was shown to be based on hidden assumptions. The truth is that the Second Law cannot be derived from the basic laws of mechanics – because, in fact, it isn't a law at all.

So what sort of a statement *is* the Second Law of Thermodynamics? Why is it useful? And does it tell us anything about the nature and direction of Time?

Let us consider for a moment a simpler 'Law' – the Law of Shuffling. This says that if you shuffle a deck of cards, the randomness of the deck either increases or stays the same. It never decreases.

Like the Second Law, the Law of Shuffling suffers from two defects: firstly, it is not absolutely true; there is always a possibility that shuffling a deck will make it less random than it was before; and secondly, the notion of randomness is not well defined. When you open a new pack, the suits are in numerical order and the suits go clubs, diamonds, hearts, spades. There are $52! \approx 10^{68}$ different ways of arranging 52 cards but there is only one way to put them in numerical order. The degree of disorder in the pack is extremely low and we say that the pack has very low 'randomness' equal to -68. (For technical reasons, entropies are always negative. To see how entropy can be calculated see **Randomness, Order, Disorder and Entropy** on page 110.) Now the pack is shuffled. Obviously (?) the pack is now very much more disordered and consequently has a much higher 'randomness'.

That's all very well, but consider this. Suppose that after shuffling the deck the cards are in the following order: 2 of spades, 7 of diamonds, 3 of clubs, Jack of clubs, etc. etc. with every one of the 52 cards specified. How probable is this outcome? Sounds pretty likely really. You might, in fact, be tempted to say that it was quite probable.

Hang on a minute! How probable is THIS outcome? If you think about it you will quickly realise that the probability of the 52 cards being in exactly this order is exactly the same as the probability of them being in numerical order – 1 in 10^{68}. This is staggeringly improbable and normally one would say that this order of improbability amounts in practice to an impossibility – and yet, every time you shuffle a deck of cards really well, an improbability of this order of magnitude happens.

So what has happened to our concept of 'randomness'? Why do we single out the numerically ordered deck as being different somehow to all the apparently far more probable arrangements which are in fact equally improbable?

The answer lies in the external context.

If you are playing bridge, the ordered deck would result in a highly unusual game. Now it is only a convention that the cards are dealt one at a time. The way a shuffled deck is dealt should make no difference so let us suppose that the cards are dealt to the four players 13 at a time; one player would end up with all the spades and would be able to declare a winning hand without playing a single trick. On the other hand, if you were playing a game of patience, the ordered deck might turn out to play like any other typical deck. There would, however, be certain orders of the cards which might, on a cursory glance inside the deck, look perfectly random but which played very strangely indeed e.g. by 'coming out' instantly.

Whether a deck of cards is counted as ordered or not depends on the game you are playing.

In the same way, the entropy of a gas in a box, unlike its energy, cannot be defined solely with reference to the positions and velocities of the molecules in the box. You also have to state what game you are playing. As soon as you have done that you can define entropy in the following way – the entropy of a system in a given state is equal to the (logarithm of the) total number of states which behave in the same way according to the rules of the game you are playing.

Now when it comes to gas molecules in a box, we are not interested in the individual motions of the molecules (the 'microstates'), we are generally only interested in the things like the temperature and pressure of the gas (the 'macrostates'). We therefore define all microstates which have the same values of temperature and pressure as being the same and we lump them all together. Because these states are so numerous, we label them as being disordered and give them a high entropy value. Any unusual state – like all the molecules travelling in the same direction, for example – will have different macroscopic properties and, because there are vastly fewer of such states, we assign to them a low entropy.

If you imagine a vast plain where each point on the plain represents a particular microstate of the gas and whose elevation represents the assigned entropy, you will appreciate that most of the plain is a high plateau; there will be occasional craters where the plain is slightly lower and occasional pits where the plain is quite low. In a minute number of places the plain will be punctured by tiny pinpricks which go down a long way but these will be incredibly rare. If you have ever seen a 3D computer realisation of the region round the Mandelbrot set you will have some idea of what I mean. Here is an image peering down into one of the abyssal depths of a fractal landscape:

(What I am asking you to imagine is technically called a 'phase space' in which each point on the plain represents a unique state of the system. Now the actual phase space of N molecules in a box has at least $6N$ dimensions, not just two – but I can't ask you to imagine that!) (For a bit more detail about phase spaces see: ***Phase Space*** on page 121.)

If you start the gas in one of the microstates represented by one of these pits and allow it to move for a time at random into any of its neighbouring states, you will appreciate that it is far more likely to move upward into the sunshine and out onto the broad plain than to dive further into the pit because any particular microstate is, in general, surrounded by huge numbers of possible states with a higher entropy – but relatively few with a lower entropy. This is why the entropy of the gas is almost bound to increase. (For more on this point see: ***Random Walks in Phase Space*** on page 122.)

You might, rightly, object at this point and say that the way a system moves from one point on the plain to another is not random at all and is governed by the laws of physics so we have no right to assume that a system will move at random from one state to the next. Absolutely correct; and this point is crucial. However, the fact remains that the laws of physics certainly *seem* to generate more and more randomness as time goes on. Gas molecules generally do not conspire to collect in one of two connected bottles; broken teapots do not often mend themselves when the bits are shaken together in a box.

Now there are two ways you can explain this behaviour. You can either maintain that a) there is absolutely no randomness in the laws of physics but the laws are constructed in such as way that they behave *as if* there is some randomness in them, or b) there genuinely *is* an element of randomness in the laws of physics.

If you adopt position a) you have a difficult task ahead of you. Indeed, this is essentially the task that defeated Boltzmann and contributed towards his suicide so I do not recommend it. In any case, it does not matter because there *is* some genuine randomness in nature and this randomness is fully described by Quantum Theory. The source of this randomness is not fully understood. My belief is that it is due ultimately to the graininess of space and time and comes about during the process of wave function collapse (see: ***Chaotic Collapse*** on page 177.)

But whatever its source, we do have an understanding of its magnitude. When two gas molecules collide, their position and momentum is not defined with absolute precision and the uncertainty is given by Heisenberg's famous relation

$$\Delta \times \Delta p \simeq h$$

118

(See: *Heisenberg's Uncertainty Principle* on page 169)

Now h (Planck's constant) is extremely small (approximately 10^{-34} Js) but when we are talking about gas molecules colliding, any tiny discrepancies rapidly multiply. Gas molecules at everyday temperatures and pressures are about 10 molecule diameters apart. Any discrepancy in position at the surface of a molecule during a collision will be magnified roughly 10 times when it makes the next collision. After 10 collisions the discrepancy will be magnified by a factor of 10^{10}. After 34 collisions, even a discrepancy as small as Heisenberg's will be magnified to macroscopic proportions. A gas molecule makes 34 collisions in about 1 tenth of a nanosecond. After this time, there is absolutely no correlation between the original positions of the gas molecules and their new positions. For all intents and purposes, the motion of the molecules is now completely random. Even the gravitational influence of a single electron at the edge of the universe will randomise the motions in less than 1 nanosecond[10].

So, to go back to the questions I posed earlier – what sort of a statement *is* the Second Law of Thermodynamics? Is is a new physical law which must be added to the list of fundamental laws like Newton's Law of Gravity and the laws of electromagnetism? Or is it a meaningless tautology?

Sir Arthur Eddington famously placed the Second Law of Thermodynamics on a pedestal far above mere physical laws like Newton's laws or Maxwell's equations. He once said:

> *If someone points out to you that your pet theory of the universe is in disagreement with Maxwell's equations – then so much the worse for Maxwell's equations. If it is found to be contradicted by observation – well these experimentalists do bungle things sometimes. But if your theory is found to be against the second law of thermodynamics I can give you no hope; there is nothing for it but to collapse in deepest humiliation.*

and he was quite right. The laws of Physics get superseded. Newton's

---

10　The gravitational attraction between an electron at the edge of the universe and a hydrogen atom is about 10^{-115} N and if the uncertainty in the position of the electron is 10^{-10} m, (the size of an atom) the *change* in gravitational force will be of the order of 10^{-160}. A gas molecule makes 160 collisions in less than a nanosecond.

law of Gravity succumbed to Relativity, Maxwell's equations to Quantum Electrodynamics; and the process will, no doubt, go on. But the very fact that the Second Law is not a law at all but a statistical rule of thumb which only requires for its proof the existence of a tiny bit of randomness in the fundamental behaviour of matter puts it way beyond the reach of any change in physical theory. Its power and scope is awesome and its utility beyond question – but *it is not the reason why time goes forwards and not backwards*. It is simply a necessary consequence of quantum randomness and the passage of time. Let us get certain things absolutely clear:

1. The Second Law of Thermodynamics cannot be *deduced* from the classical laws of physics.

2. It can, however, be deduced from the fact that *time flows* and that the laws of physics contain an element of *randomness*.

3. Although it is asymmetrical with respect to time and we can use it to figure out which way a time is flowing in a film of a sufficiently complex physical event, it emphatically does not *cause* time to flow in one direction rather than another because even if time was reversed the *same law would apply*.

While it must rank as one of the most important insights into the behaviour of matter and one of the most useful generalisations ever devised, it is not a physical law and it tells us nothing about the direction of time. It is essentially a tautology – 'in a universe which contains an element of randomness, the more ways there are in which something can happen, the more likely it is that things will happen that way'.

Phase Space

One of the most powerful ways of analysing a dynamical system is by using the concept of *phase space*. Consider a fly buzzing around a room. At any instant in time its location in the room is described by three coordinates; in addition we need another three numbers to describe its velocity. So six numbers tell us exactly where the fly is and in what direction it is moving.

Now if we had a piece of six-dimensional graph paper (ordinary graph paper only has two dimensions) we could plot all this information in the form of a single point. Since displacements and velocities vary continuously, over time the fly will trace out a continuous wiggly line in this six-dimensional *phase space*.

Now suppose that there are a dozen flies in the room. To keep track of them all you might think it was necessary to consider 12 different phase spaces but there is another way. All we have to do is define a new phase space with 72 dimensions. Now the whole system can once again be described in terms of the motion of a single point within this expanded phase space.

For those of us who find it difficult to think in 3 dimensions let alone 72 this may seem a retrograde step but mathematicians have no difficulty in analysing theorems in multi-dimensional spaces.

The phase space appropriate to the case of two connected bottles containing 1000 molecules as discussed in ***Randomness, Order, Disorder and Entropy*** on page 110 is rather unusual in that while it has 1000 dimensions, each dimension has only two possible values ('molecule in bottle A' and 'molecule in bottle B'). Nevertheless, this means that the space still contains 2^{1000} available points. Now the ease with which we can write down a number like this belies its size. The cube root of 2^{1000} is approximately 2^{333} which is equal to about 10^{100} and is quite quite a lot larger than the the number of atoms in the whole universe. So imagine a box with all the atoms in the whole universe lined up on one side completely filled with atoms from countless other universes as well and you still haven't grasped the enormity of 2^{1000}.

And this is only for 1000 molecules. What if (as is much more likely) the bottles contain 1,000,000,000,000,000,000,000 molecules?

Random Walks in Phase Space

Imagine playing the following game: you are standing halfway up a long ladder; a friend of yours tosses a coin and if the coin turns up heads you climb up one step; if it is tails you climb one step down. Various interesting questions suggest themselves.

1. On average, how far will you get after N steps?

2. What is the probability of reaching the top of the ladder in less than N steps?

3. And lastly, what is the probability that you will eventually end up a) exactly where you started? or b) at the top of the ladder?

The answers are as follows. On average after N steps you will be $\sqrt{N}$ steps away from where you started – so after 16 steps you could expect to be 4 steps above or below your starting point *on average*. (Of course this tells you nothing about where you will actually be. You might be 16 steps up, 16 steps down or anything in between!)

The answer to question 2 is not easy to calculate but there is always a finite probability of reaching the top in N steps.[11]

The real surprise comes with the answer to question 3. In the case of a 1-dimensional ladder it is *certain* that you will both get to the top of the ladder and return to your starting position given enough steps![12]

Now consider a similar game played with a King on a chess board. At each move the King moves one space N, S, E or W at random. As before, it turns out that after N moves the King will be, on average $\sqrt{N}$ steps away from where it started. Since a 2 dimensional random walk can be considered to be two independent 1D walks happening at the same time, we can still predict that the King will fall off the board eventually, however big the board is. But it does not therefore follow that the King will inevitably return to his start. For this to happen, both 1D walks would have to return to the origin *at the same time*. The astonishing thing is that, even in 2 dimensions, this is bound to happen sooner or later. Given sufficient time, our King will visit every square on the board, however big the board is!

---

11 Provided, of course, that N is greater than the helf-length of the ladder.
12 To be accurate, this is only true for a ladder which extends to infinity beyond the 'top' and 'bottom'.

But in 3 dimensions, everything changes. A fly buzzing around an infinite room *may* never return to its starting point. It might, but it might not. The probability of return is, in fact, about 34%. And if we consider random walks in higher dimensions, the probability of return gets less and less. So the probability of our original fly returning not only to its original place in the room but also with exactly the same velocity as well is equivalent to solving the return problem in 6 dimensions.

If we move up to 12 flies in a room or 1000 molecules in a bottle it is easy to see that the probability that the system will ever return to anything like its original configuration is vanishingly small. The phase space appropriate to 1000 molecules in a pair of connected bottles is vast (see: **Phase Space** on page 121.) and the chances of a random walk through this phase space returning to its original configuration with all the molecules in one bottle are virtually zero.

This theorem has an important consequence as regards the second law of thermodynamics. We have noted that the phase space appropriate to 1000 molecules in two connected bottles is truly vast. At every point in this space we can calculate a value for the entropy of the system. For example, at the origin (which describes the unique case when all the molecules are in bottle A) the entropy is -300. (For the details of this calculation see: **Randomness, Order, Disorder and Entropy** on page 110.) There are 1000 points in the phase space describing situations in which exactly one molecule is in bottle B and these have an entropy of -297. (The reason for this is that 1000 is equal to 10^3 so the entropy is 3 units higher.) and 1,000,000 points where the entropy in -294 but the vast majority (and I mean VAST majority) of points have entropy values close to zero.

Now another important to make is that neighbouring points in phase space have similar entropy values. (Mathematically speaking the entropy function is smooth rather than jumping about all over the place) What this means is that all the points which have entropy values of -297 cluster round the points with entropy value -300. In fact each of the -297 points borders on exactly 1 -300 point and 999 -294 points. If you take a random walk starting from a -297 point, you are 999 times more likely to reach a -294 point than the unique -300 point. Indeed it is easily shown that at *any* point you are much more likely to go to a point with a higher entropy than one with a lower entropy. This explains why, on average, entropy always increases.

Black Holes and Brick Walls

In the professor's first description of the edge of the universe, readers may have been reminded of a description of the event horizon of a black hole and that is exactly what I have tried to describe.

Imagine that you are suspended stationary 1 km above the surface of a black hole whose radius is also 1 km. (As we shall see, you would have to be very, very small and the rope would have to be very, very strong!)

Forgive me for indulging in a little mathematics here but the formulae are easy to use and the results rather spectacular.

As you probably know, the 'escape velocity' of a black hole at its 'surface' (the so-called event horizon) is equal to the speed of light c (= 3 × 10^8 ms^{-1}).

The escape velocity v_{esc} at any point outside a spherical mass (whether a planet or a black hole) is related to the acceleration due to gravity g at that point by the simple formula $v_{esc} = \sqrt{2\,g\,r}$ where r is the distance between the point and the centre of the mass. So by putting v equal to c and r = 1000 m, we can easily calculate[13] that the acceleration due to gravity at the event horizon of this particular black hole is 4.5 × 10^{13} ms^{-2}. Now g (which is a measure of the force of gravity) obeys an inverse square law so 2000 m away from the centre of the black hole, g will be ¼ of this i.e. about 1.1 × 10^{13} ms^{-2}. (This is about a million, million times greater than the acceleration due to gravity at the surface of the Earth so you can see why the rope holding you up must be very strong!)

Now suppose you were to poke your finger cautiously towards the black hole. Because the tip of your finger is closer to the black hole than you are, it would experience a greater force of gravity than the rest of you and would be instantly ripped off!

There is a gruesome little formula (at least when applied to fingers!) which relates the tensile stress (force per unit cross sectional area) σ in a cylindrical rod of density ρ and length l at a distance r from a black hole at the point where the acceleration due to gravity is g. It is

---

13   $g = v^2/2r = (3 × 10^8)^2/(2 × 1000) = 4.5 × 10^{13}$ m s^{-2}

$$\sigma = \frac{2\,g\,l^2}{r}$$

Plugging in the figures (digits?), we find that for a finger of length 5 cm, and cross section area 1 cm² the tensile strength needed to hold it together is about 2750 N[14] or the weight of ¼ of a ton. Ouch!

Alternatively, we could calculate the longest steel rod that could survive the pull of gravity in this region of space by making l the subject of the above equation and putting $\sigma = 6 \times 10^8$ N m⁻² (The tensile strength of steel). The answer works out to be about 23 cm[15].

The point about a black hole is that space gets more and more curved as you approach the hole. It is often likened to the shape of a rubber sheet with an extremely heavy weight placed in the middle. At the event horizon the 'slope' of space becomes so steep that not even light can crawl its way out (rather like an ant trying to climb out of a pitcher plant).

Actually, if we remain outside the hole, we cannot in all honesty say what happens inside the event horizon. That region is effectively beyond our universe and the event horizon is, in practice, the edge of the universe we live in.

The region of space near a gravitating object has positive curvature like the surface of a sphere. There is another sort of curvature – negative curvature like the surface of a saddle.

It is also conceivable that there are regions of space where the curvature becomes infinitely negative. These would be equally effective at being an edge to the universe but instead of being attracted towards such an edge, a material object would be repelled away from it and a steel rod nearby would be squashed out of existence instead of being stretched. The term 'Black Hole' gained popularity in the 1960's when it was often used in a rather tongue-in-cheek way because it was assumed at the time that such outlandish objects could not possibly actually exist. I hope I may be forgiven for introducing the term 'Brick Wall' in a similar vein to describe a place where the curvature of space becomes infinitely negative! If 'Brick Walls' are one day discovered in our

14 $\sigma = 2\,g\,l^2\,/\,r = 2 \times 1.1 \times 10^{13} \times 0.05^2\,/\,2000 = 2.75 \times 10^7$ N m⁻²
    so the tension in a finger of cross section area 10^{-4} m² is 2750 N
15 $l = \sqrt{\sigma r / 2g} = \sqrt{6 \times 10^8 \times 2000\,/\,2 \times 1.1 \times 10^{13}} = 0.23$ m

universe, I hope I may be credited with naming them first!

I would like to make one last point concerning the professor's demonstrations. In discussing relativity, much confusion is caused by authors (including myself) who fail to distinguish between what things *appear to be* and what they actually *look* like (not to say what they actually *are!*).

In Special Relativity, when a spaceship (whose length is actually 100 m in length according to the man who built it) flashes past me at half the speed of light, I may *infer* from my measurements that it has been contracted to 86 m – but it won't *look* like that to me as I watch it go by. Because of the Doppler effect, it will *look* as if it is 172 m long as it approaches and 57 m long as it recedes[16].

Things are even worse in General Relativity because things can *never* be described as they actually *are*; they are *only* what they *appear to be*. For example, when I describe a black hole as being 1 km in diameter – this is only how I infer it to be from measurements made from *outside*. Because the hole distorts the very space it is sitting in, measurement made *inside* the hole would give very different, possibly infinite answers.

As to the question of what a Black Hole would actually look like, it follows from the very definition of the object from which light cannot escape, that you cannot 'see' it at all. All you could hope to see is light being emitted from objects as they fall into the hole. Indeed it is this very light (in the form of powerful X-rays from the centre of our and other galaxies) which forms the strongest evidence we have for the existence of Black Holes. So if you were suspended 1 km above the surface of a Black hole 1 km in diameter, what would you actually *see*? In addition to the x-rays being emitted by atoms streaming past you into the hole, you would see photons from objects that fell in a week ago, microwaves from objects that fell in a year ago and even radio waves from objects that fell past you centuries ago but whose final electromagnetic cries have taken aeons to climb back out again. In short, the hole would look anything but black!

---

16 When the spaceship is approaching, the light which starts at the back of the ship is chasing the front which means that it has to start out a lot *earlier* if it is to reach my eyes at the same time as light starting from the front. This makes the ship look *longer* by factor of 2. When the ship passes me, the relative speed of the ship and the light is 1½ times the speed of light so the ship looks as if it is only ⅔ as long as it should.

One thing, however, that would be immediately obvious is that the background of stars would be seriously distorted by the gravitational bending of starlight as it passed the black hole. It would look as if all the stars behind the black hole were repelled away from it. If there was a star directly behind the hole, its light would be spread out into a ring surrounding the hole and inside the ring you would see another image of all the stars in the sky! I would love to see a black hole (from a safe distance of course!).

I really have no idea what a Brick Wall would look like. Perhaps someone could enlighten me?

General Relativity and the Shape, History and Fate of the Cosmos

General Relativity has opened up a Pandora's Box of opportunities for the speculative cosmologist. Once you have admitted that the presence of matter and radiation can distort space-time, you can start to imagine all sorts of weird universes – not just spherical or saddle-shaped universes but doughnut-shaped universes, universe with wormholes connecting one part to another and even universes in which time travel is possible.

Fortunately (or regrettably if you prefer) there is absolutely no evidence to suggest that our universe is anything other than a simply-connected 3-dimensional space and, apart from the odd Black Hole here and there, it is probably 'flat' as well. On the other hand, it definitely has an interesting history and its future is uncertain, and if we want to explore either of these aspects, we must take General Relativity into account.

The most fundamental thing which General Relativity has to tell us is that matter curves 3-dimensional space in the same way that a sphere is curved. Mathematicians call this positive curvature. If you add up the angles of a triangle on the surface of a sphere, you will find that it comes to more than 180°; if you take a piece of rope 1 m long and lay it out in a circle, you will find that the area it encloses is greater than the expected $1/4\pi$ m^2. Exactly the same things apply to space containing matter. The figure below shows a triangulation survey of the space round a Black Hole.

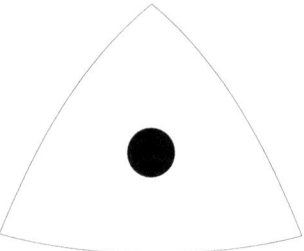

The laser beams curve their way round the hole following 'straight lines' in curved space and it is immediately apparent that the three angles shown add up to more than 180°.

Now, if we make the very plausible assumption that the average distribution of matter in space is everywhere the same (at least on a sufficiently large scale) we would have to conclude that space, like a sphere, closes in on itself and, at any point in time, is finite in volume (but has no edge). We would also have to conclude that, owing to the gravitational attraction of matter for itself, such a universe would be unstable and would collapse in on itself in a finite time. These conclusions troubled Einstein for a number of years until it was discovered (by Edwin Hubble in 1929) that the universe was actually expanding. This put a completely different complexion on things and it opened up a new range of possibilities the theoretical details of which had actually actually been worked out several years earlier by the Russian physicist Alexander Friedmann. (Sadly, Friedmann died of typhoid fever in 1925 so he never knew how valuable his contribution would turn out to be.)

Friedmann's equations are worth quoting because they are, in a way, as important as Newton's law of gravity or Maxwell' electromagnetic equations.

$$
\begin{aligned}
H &= \left(\frac{1}{3}\Lambda + \frac{8\pi G}{3}\rho - k\frac{c^2}{a^2} \right)^{1/2} \\
\dot{H} &= \frac{1}{3}\Lambda - \frac{4\pi G}{3}\left(\rho + 3\frac{P}{c^2} \right)
\end{aligned}
\qquad (1)
$$

Without going into great detail, let me explain what these equations tell us. The first one is a formula for the Hubble constant H. This tells us how fast the universe is expanding. The second equation tells us the rate at which the Hubble constant changes over time.

Two symbols need explaining first. ρ is the average density of matter in the universe at any given time and P is a measure of the amount of radiation (photons) in it.

We shall start by simplifying these equations in two ways. The funny symbol Λ (which is the upper case version of the Greek letter λ) is the famous 'cosmological constant and for the moment we shall assume that it is zero. Also we shall only consider, for the moment, 'flat' universes – i.e. universes with Euclidean spatial geometry. In these universes, k is also zero. (The exceptionally alert reader may object at this point that, since I have explained that matter causes space to bend, how can I maintain now that our universe is 'flat'. The answer is that matter does not just bend space, it bends space-time. In a static universe, the presence of matter will cause space to curve but in a dynamic universe which evolves over time, a *slice through space-time* at any instant t may be flat but space-time is still curved. This was the subtlety which eluded Einstein for so long so you may be forgiven if it escaped you as well!)

Eliminating these terms we get:

$$H = \left(\frac{8\pi G}{3}\rho\right)^{1/2}$$
$$\dot{H} = -\frac{4\pi G}{3}\left(\rho + 3\frac{P}{c^2}\right)$$

(2)

Now it should be immediately obvious from the first equation that if there is any matter in the universe at all (i.e. if $\rho > 0$) the universe *must* be expanding (H must be positive)! All *flat* universes expand forever. On the other hand, the second equation tells us that as long as there is either matter or radiation in the universe, the rate of expansion will slow down ($\dot{H}$ will be negative).

The converse of this is that, in the past, the universe must have expanded faster than it does now. This is a bit of a problem for the following reason. Using the currently accepted value for the Hubble constant, the universe cannot be older than about 9 billion years. Although twice as old as the Earth, this does not really give cosmologists enough time to explain how all the stars and galaxies and heavy elements came into being before the formation of our own star[17].

Another objection is that there does not seem to be enough matter in the universe to fit with the first equation above. Plugging in the known values of H and G (the universal gravitational constant) gives us

---

17 H is $1/13,600,000,000$ y^{-1}. What this means is that every year, the universe expands by $1/13,600,000,000^{th}$ of its current size. It follows that if the universe contracted at the same rate as it is currently expanding, it would shrink to zero size in 13.6 billion years. The Friedmann equations tells us that the universe cannot be older than $\frac{2}{3}$ of this, ie 9 billion years.

a figure of 9.7×10^{-27} kg m^{-3} or about 6 protons in every cubic metre[18]. Now it is obviously very difficult to estimate the density of matter in the universe but at the moment, our best guess as to the amount of matter we can actually see tots up to about 0.5×10^{-27} kg m^{-3} i.e. about 5% of what is called the *critical density*. There is strong evidence that matter which we cannot see (dark matter) exists in the universe, particularly in the centres of galaxies, but this is thought to account for only another 25% or so leaving a very significant shortfall. Nevertheless, it is worth noting that the critical density of the universe ρ_{crit} is related to the Hubble constant H by the equation:

$$\rho_{crit} = \frac{3\,H^2}{8\,\pi\,G} \qquad (3)$$

So perhaps the universe is not flat after all. Perhaps k is not zero. Now in a positively curved universe, k is greater than zero and in a negatively curved one, k is less than zero. Looking at the first equation in box (1) again, we see that if ρ is smaller than it should be, k must be *negative* to redress the balance. This means that our universe must have overall negative curvature.

This universe has the attractive feature that we can put just as much, or as little matter into it as we find, but it suffers from the same disadvantage as the flat universe in that it must be significantly younger than 13.6 billion years though not to quite the same extent.

Its big disadvantage is that the recent measurements on the Cosmic Microwave Background radiation (the CMB) made by the Wilkinson Microwave Anisotropy Probe have pretty well conclusively proved that the universe is flat and therefore that its density has got to be 9.7×10^{-27} kg m^{-3} after all, in spite of what we can or cannot see.

This leaves us with a dilemma. The simple way out is just to say that there must be a lot more dark matter in the universe than we think. This is certainly plausible but in the last two decades, evidence has been growing that, far from slowing down, the expansion of the universe is actually speeding up! For this, and other reasons, another idea has been gaining popularity.

Do you remember that funny symbol Λ that we chose to ignore a

---

18 $H = 1/(1.36 \times 10^{10} \times 365 \times 24 \times 60 \times 60) = 2.3 \times 10^{-18}$ s^{-1}. $G = 6.7 \times 10^{-11}$ N m^2 kg^{-2} .

while ago? This is called the 'cosmological constant'. Einstein originally introduced it so that he could construct a stable, static universe with matter in it; then he rejected it when it was discovered that the universe wasn't static after all; now it is back!

If we put $k = 0$ (and also ignore the P term which is thought to be much smaller than ρ) the Friedmann equations reduce to:

$$H = \left(\frac{1}{3}\Lambda + \frac{8\pi G}{3}\rho \right)^{1/2}$$

$$\dot{H} = \frac{1}{3}\Lambda - \frac{4\pi G}{3}\rho \tag{4}$$

Now we can make up the shortfall in density by assuming a positive value for Λ. This has an additional benefit because, providing the actual density of matter (and radiation) in the universe is less than ⅔ of the critical density, $\dot{H}$ will be zero or even positive[19]. This means that the universe will be at least as old as 13.6 billion years – perhaps even older.

The first Friedmann equation tells us that Λ acts like a kind of matter in that it adds to the total amount of matter which causes the universe to expand; on the other hand, instead of acting to slow down the rate of expansion, it actually speeds it up. In this respect Λ-matter (commonly but very confusingly called 'dark energy') is the long-looked for anti-gravity matter of science fiction!

Nobody seriously imagines that our current theories constitute the end of the affair.

---

19 If we make the substitution $\Lambda = 8\pi G \rho_\Lambda$ and put $8\pi G/3 = \mathbf{G}$, we can simplify the equations down to $H^2 = \mathbf{G}(\rho_\Lambda + \rho) = \mathbf{G}\rho_{crit}$ and $\dot{H} = \mathbf{G}(\rho_\Lambda - \rho/2)$ from which we can easily see that if $\dot{H}$ is to be positive, ρ must be less than twice ρ_Λ and therefore less than ⅔ of the critical density.

The Twins Paradox

It is well known and has been experimentally verified that moving clocks go slow. Indeed, the GPS device in your mobile phone has to take these effects into account when working out your position so the fact cannot be denied.

If, therefore, you travel around for a while in a fast spaceship and return to Earth, you will find that your watch has recorded less time than the clocks which stayed stationary. To calculate the size of the effect you can use the simple (approximate[20]) formula:

$$\text{Time difference} = \text{Time spent travelling} \times \frac{1}{2}\frac{v^2}{c^2}$$

where v is your speed and c is the speed of light (3×10^8 m s^{-1}).

For example, an airline pilot with 10,000 hours spent flying at 500 mph (222 m s^{-1}) will be younger than his stay-at-home wife by 10 μs[21].

This is not in doubt. So where lies the paradox?

The paradox lies in the following argument. Since (as Einstein himself said) all motion is relative, you could just as well argue that it was his wife who was in motion and the pilot who was stationary. In which case, the wife would be 10 μs older than the pilot. They can't both be right.

The crucial difference between the two individuals is that it is the pilot who accelerates, decelerates and turns corners while his wife stays in what is called an 'inertial frame'. (We must, of course ignore the motion of the Earth in this.) You can't argue that the pilot was stationary and it was his wife that was moving because the pilot can point to the swinging pendulum and rotating gyroscope in his cockpit to prove that his motion is changing.

---

20 The formula is only valid if v is a lot less than c

21 Time difference $= 10,000 \times 60 \times 60 \times \dfrac{1}{2}\dfrac{222^2}{(3 \times 10^8)^2} = 9.9 \times 10^{-6}$ s

Newton's Corpuscular Theory of Light

In Newton's day it was natural to think of light as being a stream of particles – after all, the most obvious thing about light is that it travels in a straight line and that's exactly what particles do (when free of all frictional or gravitational forces). Waves, on the other hand, spread out all over the place.

In addition, armed with his recently formulated laws of motion and gravity, Newton was able to put together a half decent explanation of the reflection and refraction of light off glass: reflection was caused by the particles (or 'corpuscles') bouncing off the surface like a rubber ball and refraction was due to the attraction of the material medium of the glass and the corpuscles as they approached the surface.

This is how Newton saw reflection and refraction:

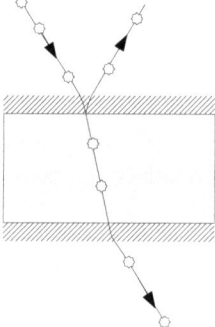

When the corpuscles enter the shaded region near the glass block, they are attracted towards the block and so the angle of the ray bends, as it does, *into* the block.

Newton had some difficulty with explaining why some corpuscles enter the block and some bounce off but he satisfied himself (and many other English scientists of the day) that his theory was correct. Nevertheless, he was well aware of the existence of the wave theory of light (Christiaan Huygens, its main exponent died in 1695, long before Newton) and it is significant that his great treatise on light – Opticks – makes very little mention of corpuscles and is equally applicable to either theory.

Huygens' Wave Theory

Huygens published his theory that light was a wave in 1690. He pointed out that both reflection and refraction had simple explanations using waves and the well known Snell/Descartes law of refraction (sin i / sin r = constant) has a particularly elegant proof if you assume that light waves travel slower in glass than in air.

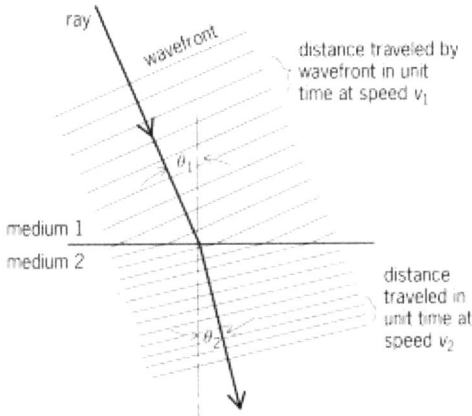

This contrasts sharply with the prediction of Newton's corpuscular theory which requires that light travels faster in glass than air.

The phenomena of interference and diffraction was little known and though the existence of colours in soap films was obvious enough, until Newton himself had clarified the nature of colour, its connection with the wavelength of light could hardly have been appreciated. Indeed, the issue was not finally resolved until the experiments of Young and others in the early 19th century.

The Photon Theory of Light

In the 1890's, one problem above all others was exercising the brains of the best physicists of the day. It is a common everyday experience that a red-hot poker emits radiation and light and that the hotter it is, the more radiation it emits and its colour changes. It is easy enough to measure the amount of radiation emitted and its colour and the experimental results can be displayed as a family of graphs.

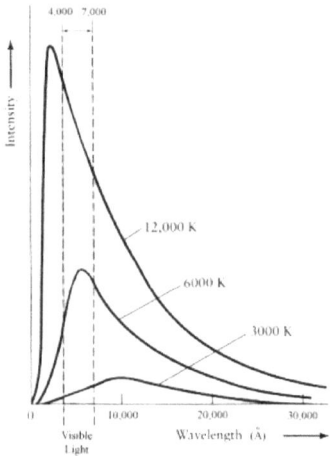

As the temperature of the object increases, the area under the curve (representing the total amount of energy being radiated) increases; in addition, the 'hump' moves towards shorter wavelengths (indicating the change in colour).

In 1896, Wilhelm Wien proposed a formula which fit the family of curves pretty well at the short wavelength end but greatly underestimated the long wavelength contribution. A few years later, Lord Rayleigh and Sir James Jeans derived a new formula using well-proven ideas from classical thermodynamics – notably the equipartition of energy – but this formula turned out to be worse than Wien's original. It fitted the facts very well at the long wavelength end but predicted that the energy radiated in the ultraviolet and X-ray region of the spectrum would increase without limit!

The breakthrough came when Max Planck had the bold idea of rejecting the equipartition of energy. In classical thermodynamics, it had been proved that in any system of freely coupled oscillators, every oscillator (or mode of oscillation) would share the available energy equally between them in the same way that a teacher, taking a party of schoolchildren to the fair one day, would share out a pile of pennies by giving each child the same number of coins.

But suppose that all the boys (who only wanted the sixpenny rides on the dodgems and had no interest in the penny-in-the-slot machines) insisted on having sixpences. No problem. The teacher gave the boys 3 sixpences and the girls 18 pennies (The available money came to 1 shilling and 6 pence for each child). All was well until the fair opened a new attraction the roller coaster! The cost for a ride was a shilling so when the children queued up for their money, the boys now insisted that they got all their money in shillings and were, naturally very disappointed when they learned that they would get less than their fair share of money.

The point of this story is as follows. If some oscillators – the high frequency oscillators in particular – insist on having large chunks of energy, they will get less of the available energy than oscillators which are prepared to accept energy in smaller chunks. Indeed, if the chunk required is too high, the oscillator may end up getting no energy at all.

This was Max Planck's big idea. Oscillators (i.e. atoms) had to emit radiation in chunks whose energy E was proportional to the frequency v of the oscillator. In short:

$$E \;=\; hv$$

where h was a constant (now called Planck's constant).

When he added this idea to Rayleigh's classical derivation, it worked out perfectly. The irony was that, at the time, Planck considered his idea to be just a mathematical dodge. He didn't really believe that light *travelled across space* in chunks, only that it was *emitted* in chunks. All that was to change when an impecunious clerk in the Patent Office in Bern published an astonishing paper in 1905 which not only assumed the reality of these chunks of light (now called 'photons', a term first used by G.N.Lewis in 1926) but used it brilliantly to explain one of the other outstanding problems which haunted scientists at the end of the 19th century – the photoelectric effect.

When you shine a strong light (a wave) onto certain surfaces, electrons (a particle) are emitted. There is, on the face of it, nothing particularly surprising about this effect. If a powerful water wave crashes onto a beach, nobody is surprised when the pebbles get jostled and moved about.

The odd thing about the photoelectric effect is this. *Blue* light works but *red* light doesn't, however strong the light! Now the only difference between blue light and red light is that the former has a slightly shorter wavelength. It is not easy to see why the *wavelength* should have anything to do with the matter. Surely it is the *amplitude* (the height of the wave) which matters?

And another thing. If you shine a strong blue light on the surface, copious electrons are emitted and if you reduce the intensity of light, fewer electrons are emitted. Again, nothing funny about that, you might say. The smaller the amplitude of the waves crashing on the beach, the less the pebbles will get jostled about. In fact, you might go on to say, if the amplitude is very small, like little ripples, there won't be enough energy around to jostle any of the pebbles at all.

Wrong. It doesn't matter how weak the blue light is, the number of electrons is strictly proportional to the intensity of light. Even the tiniest ripple will dislodge an electron eventually.

Einstein's explanation (yes, the impecunious clerk was none other than the famous author of Relativity, Albert Einstein) was breathtakingly simple:

Blue photons have a shorter wavelength than red photons; they therefore have a higher frequency and slightly more energy; a certain minimum energy is needed to knock an electron out of a metal; blue photons have just enough energy, red photons don't.

What is more, however weak the light is, when a blue photon arrives, an electron will be emitted.

What could be simpler or more convincing?

Photons had to be real.

Light is emitted in photons and light is captured in photons.

The Wave-packet Theory of Light

Light is emitted in photons and light is captured in photons. (See: **The Photon Theory of Light** on page 136) - but does it follow that light is *transmitted* in photons?

You can agree that light is emitted and absorbed in chunks but still believe that light is really a wave by introducing the idea of a wave-packet. This is nothing more than a short burst of waves like this:

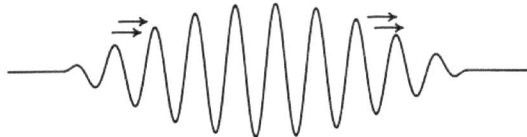

This is a very attractive idea. It seems on the face of it to combine in a very neat way the properties of both waves and particles. On the one hand, it has a (fairly) definite wavelength and it is also (pretty well) localised in space. What is more, you can relate the uncertainty in its wavelength (or momentum) to the uncertainty in its position using Heisenberg's uncertainty principle . For example, in the picture above, the wave packet is about 5 cm long and contains about 10 waves. Its wavelength λ is therefore about 0.5 cm but because of the difficulty of saying exactly where the first and last waves begin and end, we must admit to a small uncertainty in our measurement amounting to about 10% of a wavelength or 0.05 cm. In general, for a wave packet containing N waves, the uncertainty in wavelength is approximately λ/N and the uncertainty in position is $N\lambda$. The more waves the packet contains, the more accurately you can determine the wavelength but the less accurately you can determine its position.

Now in the case of a photon of light, the momentum p of a single photon is related to its wavelength by the formula:

$$p = \frac{h}{\lambda}$$

where h is, of course Planck's constant. To find the *uncertainty* in p (Δp) given the *uncertainty* in λ $(\Delta \lambda)$ we must differentiate this formula. This gives us:

$$\Delta p = - \frac{h}{\lambda^2}\Delta\lambda = - \frac{h}{\lambda^2}\frac{\lambda}{N} = - \frac{h}{\lambda N}$$

Now the *uncertainty* in the position of the wave packet (Δx) is equal to $N\lambda$, so multiplying Δp by Δx we get:

$$\Delta p \Delta x = - \frac{h}{\lambda N} \lambda N = (-)h$$

which is Heisenberg's uncertainty principle. (The minus sign is not significant)(See page 169).

(NB this is not a *proof* of the uncertainty Principle, all we are saying is that the idea of a wave packet is *consistent* with the principle and gives us a valuable insight into why the principle holds.)

Problems arise, however, when you ask the question 'How long is a photon then? If it is a wave packet, it must have an approximate length, last for a certain time and possess at certain number of wave cycles. How long? How many? Be precise.'

One of the most interesting questions you can ask about a photon is 'Over what distance will a photon interfere with itself?' If you have read *The Double-slit Experiment* on page 146 you will know that even a single photon can go through two slits at once and cause an interference pattern on a screen. Part of that pattern is due to the fact that the bit of the photon that goes through one slit has to travel further than the bit that goes through the other. We can perform a similar experiment in an instrument called an interferometer in which one bit of the photon has to travel considerably further than the other and we find that, as we gradually increase the path difference, the interference pattern gradually fades away. When the path difference reaches a certain value, the interference disappears. This is called the *coherence length* and for visible light it is of the order of a metre or so.

On the face of it, this seems to be powerful evidence that wave packets really do exist. If a photon of visible light is about 1 m long, then it lasts for about 3 ns and contains about 2 million waves. Unfortunately, it turns out that coherence length is not fixed. It depends greatly on how you measure it. For the author's explanation of coherence length see: *Coherence Length Explained* on page 188.)

Feynman's Theory of Light

Feynman's theory of light is not usually called that. It is called Quantum Electrodynamics (QED) and is the essential tool which all applied scientists use every day to work out problems in chemistry and solid state physics. It is, of course, a quantum theory and does not require or even invite interpretation. It is simply a method for getting extremely accurate answers to all questions involving the electromagnetic interaction between sub-atomic particles. (It does not address the weak or strong forces inside the nucleus)

In so far as it applies to photons, it can be used to explain how light gets from A to B and its premise is bizarre. In going from A to B a photon can go anywhere it likes!

No, that's not quite right. Let me explain.

When a light source emits a flash of light from point A, it sends out, not a photon, but a shower of spinning catherine-wheels (called *amplitude vectors*) which cavort through space in all directions, gradually decreasing in size as they go according to an inverse square law of time.

You cannot use the theory to say *where the photon will go*, but you can use the theory to determine the *probability* that the photon will appear at point B at a certain time afterwards. Lets see how it works. Here are points A and B with some of the paths that our spinning catherine-wheels might take.

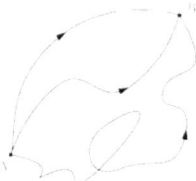

Now for each possible path, you measure its length. This tells you how big the amplitude vector of the catherine-wheel is and what direction it points in when it reaches point B.

Next you place all these vectors end-to-end (i.e. add them up together) and the (square of the) result is probability that the photon will appear at B!

If this sounds like an awful lot of work, let me reassure you; it can be simplified enormously. In the first place, paths that wander a long way off will have such small amplitude vectors by the time they get back, they can be pretty well ignored. More importantly, many vectors will be cancelled out by another vector that has approximately the same size but is pointing in the opposite direction. In fact, the only paths that contribute significantly toward the final result are those which cluster round the straight line path from A to B.

It is easy to see how this bizarre idea can generate the right results for all the wave-like properties of light – reflection, refraction, diffraction and interference etc. as well as being a *quantum* theory involving single particles and probability. However, no-one, not even Feynman, seriously asks us to actually believe in the existence of those spinning wheels. The only trouble is, the theory is too successful to ignore.

There is a certain quantity called the spin g-factor of the electron which classical theory would say must be equal to 1. A simple application of quantum theory (as developed by Dirac) suggests that it should be equal to 2 (which means that an electron is twice as efficient at producing a magnetic field as it ought to be). In effect, Dirac has just counted one way in which a spinning electron can produce a magnetic field. Feynman says ' I can think of dozens of bizarre ways in which an electron can do that and you have to take then all into account!' If you do as he says, you get the answer 2.002319304361 (or thereabouts).

This same quantity has been measured experimentally to very high precision. The result? 2.002319304199. The predicted value only begins to differ in the 9th decimal place – that's an accuracy of one part in a billion! (The discrepancy is more likely to lie with the prediction than with the experimental result because even our most powerful computers can't do all the calculations necessary to add more terms to the Feynman series.) A rifle that could fire a bullet with an accuracy of 1 part in a billion could hit a target the size of a large dinner plate on the surface of the Moon!

Surely a theory this accurate has got to have something going for it!

Superposition

It is a fundamental principle of Quantum Mechanics that a system can be in several states at the same time. Perhaps the simplest possible example of an superposed state occurs when a single photon from a source **S** impinges on a half-silvered mirror **M** as shown in the diagram below.

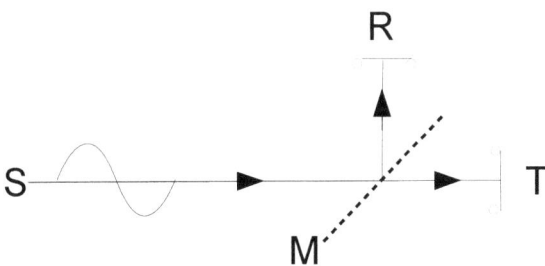

The photon has a choice. Either it passes through the mirror and is detected by the photographic plate at **T**, or it is reflected off it and detected at **R**. In a classical world, the choice would be made at random and an instant after hitting the mirror would be found *either* transmitted *or* reflected. But in a quantum world, this is not the case. Countless experiments have proved without doubt that the photon enters a state in which it is *both* transmitted *and* reflected! It is true that, eventually, when the photon is actually detected by one of the photographic plates, it turns up in only one of the two places but, for a while at any rate, it really does seem to be in two places at once. Erwin Schrödinger, one of the founders of modern Quantum Theory tried to ridicule this idea by suggesting that, if this were true, a cat could be both dead and alive at the same time (see *Schrödinger's Cat* on page 145) but his bluff was called and this is exactly what some interpretations of Quantum Theory affirm.

A more attractive possibility (to me at any rate) is that when a photon (or a cat) enters a superposed state, its state is temporarily *undecided*. It has the possibility of being in either state but Nature has not yet worked out which state it is going to be. This becomes apparent sometime later when the particle (or object) has interacted with a sufficiently large amount of its environment (a process known as *decoherence* or *objective reduction*). What causes this collapse (if, indeed, it actually occurs) is unknown. It might be something to do with the quantum nature of gravity (as Sir Roger Penrose believes) but I prefer to think, along with many others, that it is brought about by the ever increasing complexity of its relation with the environment with which the system inevitably comes into contact. (See **The Collapse of the Wave Function** on page 153)

Schrödinger's Cat

There can be few people who have not heard of Schrödinger's infamous Cat and it is not necessary to discuss the hypothetical experiment in detail[22]. Suffice it to say that when Erwin Schrödinger heard about Niels Bohr's suggestion that photons could be in two places at once he tried to disprove the idea by extending it to the point of ridicule, by suggesting the possibility of a cat being both dead and alive at the same time.

To which Bohr replied: "Why not?

"You can't prove that the cat is either dead or alive until you open the box and when you do, I predict with certainty that there is a 50% probability of the cat being alive and a 50% possibility that it is dead."

With which, of course, Schrödinger could not argue.

My take on the issue is that it is virtually impossible to place an object as large as a cat at room temperature into suspended reality because of the huge number of particles involved in random motion. On the other hand, my theory of consciousness requires pretty well just that so time will tell.

In March 2010 it was reported that "Andrew Cleland at the University of California, Santa Barbara, and his team cooled a tiny metal paddle until it reached its quantum mechanical 'ground state' — the lowest-energy state permitted by quantum mechanics. They then used the weird rules of quantum mechanics to simultaneously set the paddle moving while leaving it standing still. The experiment shows that the principles of quantum mechanics can apply to everyday objects as well as as atomic-scale particles." which gives you some idea of the distance we have yet to go – but it also shows that the idea of a cat being in two states at once is not as ludicrous as Schrödinger supposed.

---

22 Whenever I refer to the Schrödinger's cat experiment, the one I have in mind is where a single photon is directed at a half silvered mirror. If the photon passes through the mirror, it impinges on a photocell which triggers a gun and kills the cat. When the photon is reflected, the cat survives. The probability of the photon being reflected is assumed to be 50%.

The Double-slit Experiment

Richard Feynman was fond of saying that all of quantum mechanics could be gleaned from carefully thinking through the implications of this single experiment so if you are not already familiar with it, I will try to explain it as simply as possible.

Light is a wave.

If two waves arrive at the same place at the same time they add together in a rather special way. If the crests and troughs of the two waves coincide, the resultant wave is twice as big as the first one. (This is called *constructive interference*) If, however, the crests of one wave coincide with the troughs of the other they interfere *destructively* and the result is – no wave at all.

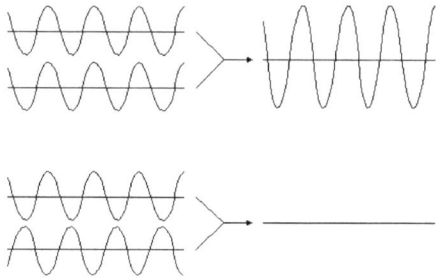

Even before we start to talk about quantum theory, this effect is quite remarkable. Two waves adding up to make one wave twice as big seems reasonable enough – but how can two waves, each presumably carrying energy, *cancel* each other out? What happens to the cancelled energy? Does it just disappear?[23] Even so, they do.

You have probably already experienced this effect yourself anyway. Radiowaves are (of course) waves too and you may well have found that your transistor radio doesn't work very well on the windowsill, but move it a couple of feet sideways and it works fine. What is happening is that there is probably a reflection from a building nearby which means that where you live there is not one signal but two – and at the windowsill they interfere destructively so the radio will not

---

23   The answer to this little riddle is that a wave which is twice as big actually has *four* times as much energy and whenever two waves cancel out, there is always another pair which are adding up. In the diagram above, the 4 waves on the left contain $1^2 + 1^2 + 1^2 + 1^2 = 4$ uinits of energy. The two on the right contain $2^2 + 0^2 = 4$ units as well.

work.

Now there are many ways of splitting a light wave into two and bringing the two beams back together again but the simplest is by using a double slit. This is simply a piece of stiff cardboard with two narrow parallel slits cut into it. You can make one yourself quite easily with a craft knife. Make sure that you can see light through both slits and that they are no more than 1 mm apart.

Now, when it is dark, hold the slits vertically in front of your eye and look through them at a distant street lamp (preferably a sodium lamp). You should see something like this:

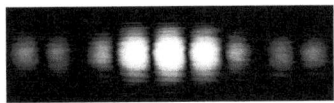

Instead of a single blob of light, you will see several blobs spread outside sideways with completely dark patches in between. It is here that waves which pass through the left hand slit interfere destructively with waves passing through the right hand slit. It is not difficult to see that you could put such a slit in front of a camera and capture the image above on photographic film.

- - -

Excuse me, light is not a wave, it is a stream of particles.

When light hits a photographic film, the energy in the light causes chemical changes to occur in the photosensitive emulsion which ultimately causes a tiny crystal of metallic silver to appear. There is no problem with this if the light is strong enough but what if the source of light is very very weak? Surely there will come a point when to light is so weak, there is never enough energy in any one place to initiate the chemical reaction and the photographic plate will just be blank?

Not so. Even with the dimmest light, the plate still responds and the crystals of silver appear, one by one, first here, then there until the picture is eventually complete.

And why does this happen? Like I said – it is because light is not a wave, it is a stream of particles. These particles are called *photons* and they each carry a bundle of energy, each one having enough energy to cause a single chemical reaction in the emulsion.

Now when you do a double-slit experiment, some photons go through one slit while some go through other. When two photons arrive at the same place on the film at the same time, they interfere with each other according to how far they each have travelled producing the interference pattern which we see.

- - -

So which is it? A wave or a stream of particles?

To test the issue, let us set up a double-slit experiment but use such a weak source of light that we can be sure that only one photon passes through the apparatus at any one time. If light is a wave, interference will still occur but the photographic film will be blank because the waves are just too weak to cause any chemical changes in the emulsion. If light is a stream of particles, the film will be nicely exposed but since it is obviously impossible for a single particle to go through both slits at once, there won't be any interference and all we will see is a single blob of light,

It takes a week for us to be sure that enough light has passed through the apparatus for us to be sure that, if light is a stream of particles, it will show up on the film and as we wait for the expected single blob to appear on the paper in the developing tank – or not, as the case may be, we lay our bets.

In the event, we both lose. The paper shows us exactly the same image as the one we got before complete with interference bands!

You can – even more easily – do exactly the same experiment with electrons because, using a fluorescent screen, you can actually see the electrons arriving one by one. At first they appear to arrive at random but as the number of electrons increases, you begin to make out the dark bands where the electrons are not allowed to go. Now, unlike photons which have no mass and constantly whizz around at the speed of light, electrons are really perfectly ordinary bits of matter and there is really no doubt that they go through one slit or the other – so how come they interfere with themselves like waves?

And there you have it in a nutshell. If you can explain how the double-slit experiment works, you can explain anything!

Waves and Particles

The famous double-slit experiment (See: ***The Double-slit Experiment*** on page 146) throws the fundamental issue of Quantum Theory into sharp relief. How can an object like an atom, an electron or a photon be both a particle and a wave at the same time? Here are a number of suggestions.

Niels Bohr: It is a nonsense question. If you do a wave experiment, you will see a wave; if you do a particle experiment, you will detect a particle. There is no point is looking any further. What we call an electron or a photon is not like a football or a sound wave; it is a different kind of entity with its own raft of properties, some of which are analogous to the properties of footballs, some of which are analogous to the properties of sound waves. Period. (See: ***The Copenhagen Interpretation*** on page 161)

Erwin Schrödinger: Particles don't actually exist; only my probability waves exist. What we see as a particle, an electron, say, is really just a small, localised bundle of waves called a wave-packet. If you wiggle the end of a slinky spring, you can see the wave packet travelling down the spring. It can even reflect off the end or knock something over, just like a particle. (See: ***Schrödinger's Wave Equation*** on page 151)

Niels Bohr: That may be true in one dimension but your theory goes up the spout in three dimensions. In 3D, waves just go on spreading themselves out more and more. Particles don't do that.

David Bohm: I agree with Niels here; it is the particles that are really real. They behave like waves sometimes because they are guided along their paths by a pilot wave which tells them where to go and where not to go. (See: ***Pilot Waves and Virtual Particles*** on page 163)

Niels Bohr: You haven't got it, have you? The particles are not 'guided' by Schrödinger's wave. They simply don't *exist* until they hit something and then you can use Schrödinger's fancy wave equation to calculate where they will end up if you like – but even then, there are other ways to do the calculation.

Hugh Everett: That's rubbish. Particles can't just not exist and then magically appear somewhere. I agree with Erwin in that the wave aspect is more important then the particle aspect but since the wavefunction which describes any real physical quantum system – such as me here – is constantly splitting into multiple slightly different copies all the time, it is impossible to say that any of them are really real. Only the reality that the real me exists in is really real – for me at any rate. Really. (See: *The Many-Worlds Interpretation* on page 162)

To my mind, the Suspended Reality interpretation gives equal weight to both points of view. Fundamentally, entities like electrons and photons are particles which have definite properties such as position, speed, momentum and energy. But during a period of suspended reality these properties have *not yet been decided*. They are only determined in retrospect when the wavefunction collapses. It could be said, therefore that my interpretation owes a lot to the Pilot Wave interpretation but this would be a mistake. The Pilot wave interpretation is essentially non-local; the behaviour of one particle is directly affected by the behaviour of its entangled partner.

Having said that, I have to point out that there is a sense in which reality is in a constant state of bing suspended. When the photon hits a detector and causes the wavefunction to collapse, it also triggers a whole new set of possibilities whose precise trajectories have yet to be decided. So reality staggers from one uncertain future to another, never ever actually resolving itself completely. Viewed from this point of view, it could be argued that the only reality is the wave function.

Schrödinger's Wave Equation

In Quantum Theory, the position of a particle such as a photon in a double-slit experiment, cannot be described simply by a trio of numbers x, y and z specifying its exact position in 3 dimensional space at a given time. Instead, it is described by a function $\psi(x, y, z)$ which essentially specifies the *probability* of finding the photon at that place *if you were to place a detector there.*

Since the photon is moving, $\psi(x, y, z)$ must change as time goes by. The way it changes is described by a *differential equation* which, for a single, slow moving particle looks like this:

$$i\hbar\frac{\partial \psi}{\partial t} = -\frac{\hbar^2}{2m}\left(\frac{\partial^2 \psi}{\partial x^2} + \frac{\partial^2 \psi}{\partial y^2} + \frac{\partial^2 \psi}{\partial z^2}\right) + V\psi$$

At first glance, this equation looks horrendously complicated but in fact it is not very different from the equation which describes waves in 3 dimensional space, the main differences being that ψ is not just an ordinary (i.e. real) number, it is *complex* (i.e. has an imaginary component) and that the left hand side, instead of being the *second* partial differential with respect to time, it is the *first* partial differential.

In spite of these differences, we can solve this equation for a wide range of different situations in exactly the same way that we would solve the equivalent wave equation for, say, sound waves in air. That is to say, given some sort of disturbance – a loud bang for example – we could calculate exactly where the sound waves would go as far in the future as we like. Schrödinger's wave equation is exactly the same. Once you have specified how to start it, its subsequent development is completely determined. There is absolutely no randomness in it at all. We can go even further. The equation allows to extrapolate back into the past with equal ease. If we know the state of the wave function at this instant, we can also work out how it got there.

What this means is that, any interpretation of QT which denies the 'collapse' of the wave function (See: *The Collapse of the Wave Function* on page 153. See also: *The Many-Worlds Interpretation* on page 162) the complete history *and future* of the universe is already mapped out and our expectation of *free will* is just an illusion.

Schrödinger's wave equation is rather different from the standard wave equation which describes sound waves in air or vibrations in a bell but in one important respect it is very similar and that is – the equation is *linear.* (i.e. it does not contain any terms in ψ^2)

This has a very important consequence. It is this.

If $\psi = F(x, y, z)$ is a solution to the equation (e.g. if the photon can travel through one slit) and if $\psi = G(x, y, z)$ is a solution to the equation (e.g. if the photon can travel through the other slit) then $\psi = F(x, y, z) + G(x, y, z)$ is also a solution to the equation (e.g. the photon can travel through both slits) . This is called *superposition.*

There really is nothing surprising about this mathematically. The equation which describes the behaviour of vibrational waves in metal permits a bell to vibrate in any number of different ways at different frequencies (called harmonics). It also permits the bell to vibrate in several different ways at the same time and when the bell is struck, the trained ear can hear several vibrational modes simultaneously.

So when a Quantum Physicist claims that the photon is in two places at once (see **Superposition** on page 143) or that Schrödinger's cat is both alive and dead at the same time (see ***Schrödinger's Cat*** on page 145), all he is saying is that the wave which describes the situation is vibrating in two modes simultaneously.

The Collapse of the Wave Function

We have seen how an experimental set-up can be completely described by a certain mathematical entity called the wave function ψ. which evolves over time according to Schrödinger's wave equation (See: *Schrödinger's Wave Equation* on page 151). Obviously this explains very nicely how photons in a double-slit experiment can interfere with each other (see *The Double-slit Experiment* on page 146) and how electrons can be smeared out over several atoms in a molecule but it does not explain how an alpha particle can cause a visible flash of light at a particular place on a scintillation screen or how a photon from a distant star can affect a single photosensitive element in the CCD inside a camera.

According to orthodox quantum theories, at some point the wave function 'collapses' but what this means or when it happens is obscure.

According to the standard interpretation of QT (See: *The Copenhagen Interpretation* on page 161) the wave function ψ for an atomic particle is a measure[24] of the probability of *finding* the particle *if you were to place a detector* at the point in question. It is assumed that the act of detecting the particle somehow causes the wave function to collapse to a state in which ψ is essentially equal to 1 at the detector and zero everywhere else. The problems with this interpretation arise when you try to define what exactly constitutes a detector. Presumably a human eye counts; also a CCD in a camera; also a scintillation screen; but what about a single atom which emits a flash of light when the particle hits it? Does that count? Isn't that all a scintillation screen is anyway – an array of atoms that emit light when they are struck by a particle?

Worse is to come.

The wave function can collapse when the particle is *not* detected! In a classic variation on the double-slit experiment, a detector is placed close to one of the slits to determine which of the slits the electron (or whatever) actually passes through. It is found that, as soon as you switch the detector on, the interference pattern on the screen disappears – that is to say, electrons land uniformly over the whole screen. This is true not only of the electrons which pass the detector and register their presence,

---

24 Technically speaking the probability is the square root of the modulus of ψ^2

it is also true of the electrons which 'go through the other slit'. I am not kidding. The experiment has been done!

So it is not the action of detecting the electron which collapses the wave function, it is the mere presence of the detector! The policeman does not actually have to catch the thief in the act, his presence alone will prevent the crime! (So – does the detector actually have to work? Would a blow-up model of a policemen do just as well as a real one? I will leave you to ponder on that one.)

Assuming that we agree on what constitutes a detector, we can resolve the issue of the collapse of the wave function when the detector remains silent as follows. When the electron passes through the double slit, its *wave function* description is affected by both slits *and by the presence of the detector* as well. It may be that the wave function chooses to collapse into a state which causes the detector to register a particle, or it may be that it collapses into a state in which the detector remains silent. Either way, the wave function is irrevocably altered and the interference pattern on the screen will be destroyed.

The importance of the wave function collapse is that it is the point at which the *probabilities* described by the wave equation are tuned into *actualities*. Without it, it is difficult to see why we should bother with probabilities at all. If, as in the **Many Worlds Interpretation** (See page 162) the collapse never occurs and a photon encounters a 10% silvered mirror, why should transmission occur in 9 worlds and reflection in only 1? Why bother with 9 identical worlds anyway?

But if wave function collapse is a physical reality, what causes it?

The Measurement Problem

In practice Quantum Mechanics works like this.

You start with a (necessarily partial) description of an experimental set-up – a photon travelling towards a double-slit, say or a radioactive atom. You convert this description into a mathematical quantity called a wave-function. There are strict rules as to how this wave-function evolves in time and a number of different mathematical ways of getting at the answer but whichever method we use, we can determine with absolute precision what the wave function will look like a moment later. Let us choose the moment when the electron is expected to arrive at the fluorescent screen, or a moment when we think the radioactive atom may have decayed.

In order to calculate where the electron will land or whether the atom will decay we calculate how the wave-function develops. Now at any point and at any instant, there are two things we can say about a wave. Its *amplitude* (i.e. how big it is) and its *phase* (i.e. where it is in its cycle). Normally both these pieces of information are vital but at this point we do a very strange thing. We throw out the phase information and square the amplitude. What we are left with is simply an array of numbers which are interpreted as the *probability* of the electron being at the point in question, or of the atom being in the expected state.

In the case of the double-slit experiment, the procedure tells us only where the electron may and may not land – it does not tell us exactly *where* it will land. In the case of the radioactive atom, we can deduce the probability of the atom decaying but not *whether* it will decay or not. In spite of this – when we see the flash of light on the fluorescent screen or hear the Geiger counter click, we know for sure where the electron has landed and whether the atom decayed.

The problem is this. In principle, the wave-function can go on developing for ever. What determines the exact point at which the wave-function 'collapses' and the array of different probabilities becomes a single certainty?

Niels Bohr suggested that the collapse occurs when a 'measurement' is made (See: ***The Copenhagen Interpretation*** on page 161) but this raises more questions than it answers.

Hugh Everett maintained that the wavefunction never collapses (See: *The Many Worlds Interpretation* on page 162) but this view has been criticized for being too profligate with cosmic real-estate!

At one point it was suggested that only a (human?) consciousness could collapse the wave function.

For my own explanation of the effect see: *The Measurement Problem Explained* on page 184

Entanglement

Photons can be polarized by passing it through a polaroid filter – e.g. a pair of polaroid sunglasses. These photons have a definite and measurable direction associated with them. A vertically polarized photon will pass through any number of subsequent vertical polarizers but it will not pass through a horizontal one. Likewise a horizontally polarized photon will pass through a horizontal filter but not a vertical one. If, however, the angle between the polarizer and the photon is 45° then 50% of the photons will pass through the polarizer and 50% will be absorbed. In fact, classical theory decrees that the intensity of the light passing through two polaroids crossed at an angle θ will be proportional to $\cos^2\theta$.

A number of experiments have been performed on what are called 'entangled' particles. In one experiment[25], two photons are sent on their way in opposite directions as a result of the decay of an elementary particle. It is a feature of these photons that their planes of polarization are always *parallel* but the exact orientation is random.

Suppose that two observers A and B set up detectors to observe the photons. Each detector consists of a polaroid filter which can be set at any angle and a photomultiplier tube which gives a loud click if the photon gets through the filter.

If polarizer A is set vertical and polarizer B horizontal, as in the illustration above, we would normally expect a small proportion of the photons (e.g. those that set out at 45° to the vertical) to pass through both polarizers but, surprisingly, this *never* happens. If detector A fires, B is silent. If detector A is silent, B fires.

Nothing wrong with that, you might say. Obviously, when the

25   This experiment is usually done with circularly polarized photons or using the spin of electrons and positrons and angles other than 45°. For more detail see: ***Bell's Theorem and the Aspect Experiment*** on page 165.

photons start out, they are both polarized either vertically or horizontally so, clearly when they arrive at the detectors, they will give the expected results.

But is this obviously correct? What if A and B set their polaroids at 45° (but still at right angles)?

The astonishing thing is that *exactly the same thing happens*. *Either* A's counter clicks *or* B's but *never* both. It seems that the two photons set out polarized vertically, say, but when the first photon reaches A's detector and encounters a 45° filter, it sends an instantaneous message back to his pal saying "Hey, watch out! I have just come across a 45° polaroid and – wheeeeey! - I have just got through it! You had better change your polarization so that if B has set his filter at 45° too, you get through too." (The only other possibility is that the photons know in advance which way A and B have set (or are going to set!) their filters but this seems even more unlikely.)

The idea that two photons could send messages to each other instantaneously was anathema to Einstein who ridiculed it as 'spooky action at a distance' and yet, if you insist that the two photons have a definite direction of polarization when they are emitted, how can you explain the results of the experiment otherwise?

Lets explore this experiment a little further.

Quantum theory predicts, and experiment confirms that if A and B are set an angle θ apart, the proportion of photons which pass through both polarizers and cause the detectors to fire together is equal to $\cos^2\theta$.

Let us suppose that, when the two entangled photons set out, they carry with them a set of identical instructions which tell them exactly how to behave when they encounter a polarizer. The simplest set of instructions might be to pass through any polarizer which is more vertical then horizontal (i.e. whose angles lie between –45° and +45° and to be absorbed by any other angle. We can represent this situation by the following diagram where light grey represents angles which are transmitted and dark grey absorbed.:

It is immediately obvious that if the two polarizers A and B are set to the same angle, the two photons will do the same thing and that if the polarizers are set at right angles, the photons will do the opposite, which is just as quantum theory predicts.

But we would soon be able to discover that this was the case because by setting both polarizers vertical we would always get double transmission. Since we find that under these condition the photons are sometimes transmitted and sometimes blocked, we must assume that the quadrature pattern can be oriented at any angle at random.

Now if the two polarizers are set at 45° then sometimes the photons will do the same (if the two angles happen to fall in the same quadrant) and sometimes they will do the opposite. In fact, it is easy to see that they must do the same exactly 50% of the time. This also agrees with quantum theory because $\cos^2(45) = \frac{1}{2}$.

But what about other angles?

Suppose we set the polarizers 30° apart. Imagine we have a pair of pointers which we can fix 30° apart and rotate around the whole circle. In what proportion of cases will we find that the segment so defined contains two colours?

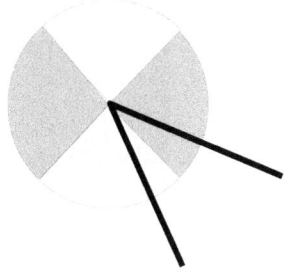

The answer is pretty obviously ⅓. What this means is that the detectors will fire together in 67% of the cases.

The problem is that quantum theory predicts, and experiment confirms, that the number of coincidences will be $\cos^2(30) = ¾ = 75\%$.

In general, the idea that the photons can decide beforehand which angles to pass through and which to reject leads to the prediction that the relation between the angle between the polarizers and the proportion of coincidences should be linear, whereas the actual relation is a $\cos^2$ curve.

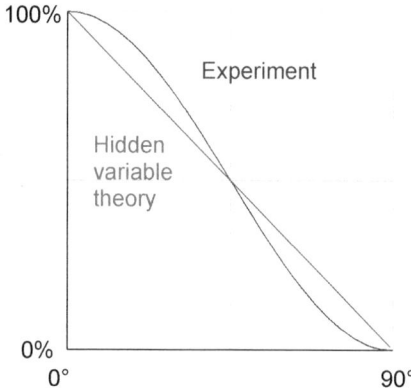

The discrepancy may appear small but it is vitally significant. When the polarizers are somewhere between 0° and 45°, a significantly greater number of coincidences are detected than can be explained by any 'hidden variable' theory. Likewise, when the polarizers are somewhere between 45° and 90°, a significantly smaller number of coincidences are detected.

It seems that we must accept one or both of the following options: either the two photons communicate with each other (which violates the assumption of *locality*) or quantum objects possess properties which have no meaning until they are actually measured (which violates the assumption of *realism*).

For my own explanation of the effect see: ***Entanglement Explained*** on page 185.

The Copenhagen Interpretation

The Wikipedia article on the Copenhagen interpretation lists the following as being the principle assumptions on which the Copenhagen Interpretation of Quantum Theory is based.

A system is completely described by a wave function ψ, representing the state of the system, which grows gradually with time but, upon measurement, collapses suddenly to its original size.

The description of nature is essentially probabilistic, with the probability of an event related to the square of the amplitude of the wave function. (The Born rule, after Max Born)

It is not possible to know the value of all the properties of the system at the same time; those properties that are not known exactly must be described by probabilities. (Heisenberg's uncertainty principle)

Matter exhibits a wave/particle duality. An experiment can show the particle-like properties of matter, or the wave-like properties; in some experiments both of these complementary viewpoints must be invoked to explain the results, according to the complementarity principle of Niels Bohr.

Measuring devices are essentially classical devices, and measure only classical properties such as position and momentum.

The quantum mechanical description of large systems will closely approximate the classical description. (This is the correspondence principle of Bohr and Heisenberg.)

It is probable that the majority of physicists who use Quantum Theory every day in their working lives use the Copenhagen Interpretation consciously or unconsciously to enable them to get results.

On the other hand, almost every one of the above paragraphs raises a question. What constitutes a 'measurement'? What definition of probability is assumed? Do objects actually have properties or not? When does a 'quantum system' become a 'classical device'? etc. etc.

The Many-Worlds Interpretation

The Many Worlds Interpretation differs from the Copenhagen Interpretation is one respect only:

The wave function never collapses.

Yes – that's it. Adherents of the MWI point to its extreme economy of assumptions and you have to admit, they have got a point.

So where does the 'Many Worlds' idea come in?

In MWI, entities like photons and electrons can exist in a superposition of states. For example, when a photon arrives at a half-silvered mirror the ensuing wave function describes a superposition of two equally possible descriptions (see ***Superposition*** on page 143.) At this stage there is always the possibility that the reflected and transmitted beams can be brought back together again and 'interfere' with each other. But when the disturbance in the wave function (notice I do not say when the *photon*) arrives at a detector, the wavefunction expands to include the detector as well and, by a process called decoherence, the two possible 'worlds' diverge and separate. In one 'world' the photon is detected at **T**, in the other 'world' it arrives at **R**. Once decoherence has happened, the two worlds can no longer interfere with each other.

In the language of the Shrödinger's Cat paradox (see: ***Schrödinger's Cat*** on page 145) it is not that the wavefunction 'collapses' into a single reality containing either an alive or a dead cat, but that reality splits into two with an alive cat in one 'world' and a dead cat in the other.

The MWI sidesteps the measurement problem completely (because the wavefunction is never collapsed) (See: ***The Measurement Problem*** on page 155) but, to my mind, it replaces it with an equally difficult problem: what brings about this process called decoherence? It is also difficult to interpret how different worlds can have different probabilities associated with them – for example, if a photon encounters a partially silvered mirror with a 10% chance of being reflected and a 90% chance of being transmitted.

Pilot Waves and Virtual Particles

The pilot wave interpretation goes right back to the early days of Quantum Theory but it was steamrollered by Bohr and his followers in Copenhagen. It is now undergoing a resurgence of popularity because it claims to solve the three central mysteries of quantum in the following way.

Entities like photons and electron are real objects existing in time and space with definite properties like position, energy and momentum but which also have a wave-function attached to them rather like the wave attached to a boat. As the particle travels through space, the associated wave-function develops in accordance with the usual quantum rules and is therefore affected by the presence of other objects nearby such as other slits or measuring devices. The trajectory of the particle is instantly guided by the wave-function so that if, for example, the other slit of a double slit is covered up, the photon will instantly 'know' that the interference pattern has been changed.

Since all the particles have real positions at all times, the question of superposition of states is eliminated; likewise there is no measurement problem either because since all our instruments, like the particles, have real states at all times, we do not have to appeal to the random collapse of a wave-function to explain them; and paradoxes associated with entanglement are sidestepped completely because the pilot wave theory is explicitly non-local and has 'spooky action at a distance' built into its very structure.

Critics of the theory have either maintained that it is just a dressed up version of the Copenhagen interpretation or that it is really a Many-Worlds interpretation in disguise because the wave-function associated with the whole universe never collapses and so contains within it information about every possible future.

Another possibility is that real particles are surrounded by a 'cloud' of virtual ones – or rather, a particle is really just a cloud of virtual ones – which run about all over the place, probing the surroundings and interacting with it. The real particle turns up in the place where all the 'virtual' particles congregate in maximum numbers. This idea owes a lot to Feynmans theory of light which instead of describing light as a wave, uses virtual photons instead. (See:

Feynman's theory of Light on page 141) The idea of 'virtual' particles has proved to be extraordinarily useful in practice but whether we can really say they exist or not is another matter.

It is possible to show that both the 'pilot wave' and 'virtual particle' ideas result in exactly the same experimental predictions so, at the moment at any rate, we have no reason other than mathematical convenience to prefer one over the other.

Bell's Theorem and the Aspect Experiment

In 1935 Einstein, Podolsky and Rosen published a paper which argued that Quantum Theory as it was then understood was incomplete. They proposed a thought experiment in which measurements were made on different parts of a quantum system (e.g. on two photons which had interacted and separated). It seemed that a measurement on one part would instantly affect the other. For example, if you measured the exact position of one of the particles, it would instantly become impossible to measure the exact momentum of the other. (If you could, you would know the exact position and momentum of both particles in contradiction with Heisenberg's Uncertainty Principle). Either the particles communicated with each other instantly (a possibility which Einstein naturally rejected) or something else was going on – the photons carried more information with them than just their energy.

The argument raged for several years until 1964 when John Stewart Bell devised a theorem which seemed to rule out the possibility that the photons could ever carry with them the information needed to produce the results which Quantum Theory dictated. Bell's theorem applies to a wide range of phenomena in which two particles are 'entangled' together and in 1974 it became possible to test his theorem experimentally. This test is essentially the one I describe in **Entanglement** on page 157 and is known as the CH74 experiment because it was carried out by Clauser and Horne in 1974.

Even more famous is an experiment carried out by Alain Aspect in 1982 shown below:

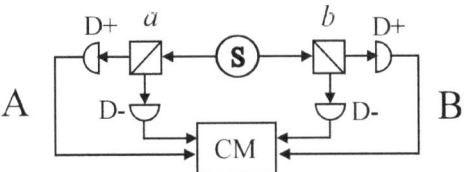

S is a source of oppositely polarized photons (they are actually *circularly* polarized); *a* and *b* are polaroid splitters which separate the photons vertically and horizontally; the four **D**'s are detectors and **CM** is a coincidence counter.

The experiment is run many times with different orientations of the two polarizers *a* and *b*. Classical and quantum theories give slightly different predictions for the number of coincidences that should be detected and the experimental results came down decisively in favour of quantum theory.

In theory, it is possible to do the same sort of experiment with electrons and positrons emitted by the decay of a pion but I do not know if these experiments have actually been carried out. Nevertheless, many different tests of Bell's theorem have been made and while it is not quite true to say that the issue is completely resolved, it certainly is the case that the great majority of physicists believe that quantum theory is king and that we have to accept one of two unpalatable consequences. Either:

a) 'spooky action at a distance' really exists – or

b) you cannot talk about the photons being 'real' until you measure them.

It is said that these experiments force us either to deny 'localism' or 'realism'; one or the other has to go.

Option a) is rejected by most physicists, not because it conflicts with Special Relativity (it doesn't) but more because nobody has the first idea how information could possibly be communicated from one photon to another in the required way. The standard approach is to abandon realism and accept that, for example, a photon simply does not have a direction of polarization until it has been measured by a 'classical' device (See: *The Copenhagen Interpretation* on page 161)

You will find my preferred explanation at: *Entanglement explained* on page 185.

The Twin Monkey Paradox

Bell's theorem predicts the results of an experiment which highlights the way in which the behaviour of quantum systems seem to violate common sense and any explanation which is based on local hidden variables. The details of the experiment are a bit technical so here is a simple parable which mirrors one of the standard experimental test of Bell's theorem in a more comprehensible way.

At the European Centre for Primate Research (CERP) near Geneva they are studying a newly discovered species of monkey. New born infants are always born twins and are initially very fussy about what fruits they like. If offered one of a banana, an orange or an apple, sometimes the new born monkey will accept it and at other times the monkey will reject it. Their fussy habits don't last long, however, and by the second day they will eat just about anything.

Now it was soon realised that if the twins were offered the *same* fruit, then both monkeys would either both accept it or both reject it even if they were housed in separate cages. On the other hand, if they were offered *different* fruits, their preferences seemed random. In fact, it seemed obvious that the monkeys carried three genes which determined their preferences for each of the three fruits, and that, naturally, twins carried the same genes. So for example, a monkey carrying genes ARR would accept a banana but reject both an orange and an apple.

For a long time, nobody thought any more about the issue but then a graduate student called Clanger noticed something very peculiar indeed. If the elder monkey of a pair of twins was offered a banana, say, and *rejected* it, his twin brother would, naturally, also reject a banana but always *accept* the fruit offered whether it was an orange or an apple. If, on the other hand, the elder monkey *accepted* the banana, the second would, naturally accept a banana too but would always *reject* the fruit offered if it was an orange or an apple. The results can be summarised as follows:

1. When two (separated) twin monkeys are offered the *same* fruit, they always make the *same* choice.

2. When two (separated) monkeys are offered *different* fruits, they always make a *different* choice[26].

Now this may not seem problematical to you – but it is.

Let us suppose the twins carry the genes ARR. Naturally, if they are offered the same fruits they will make the same choices. If the first monkey is offered a banana and the second an orange, they will make different choices; if they are offered a banana and an apple, they will make different choices; but if one is offered an orange and the other an apple, they will both reject them which contradicts statement two. In fact, whatever combination of genes the twins share, there will always be some situations in which the monkeys make choices which are inconsistent with the genetic information.

Exactly he same is true of the particles in Bell's experiment. It turns out that there is no way that the pair of particles can share common information which will allow them to behave correctly in all possible subsequent scenarios which the experimenters can set up.

It seems that we must accept that, either the particles can communicate with each other instantly over great distances or they can see into the future. Neither option is very appealing.

If monkeys were quantum particles then the standard way of explaining this curious behaviour would be to say that the two monkeys form a single system and that their preferences are not defined until they are measured – but to my mind that does not really solve the problem. You still have to explain how it is that when one monkey's preferences are revealed, how is that information communicated to the other?

(For my preferred explanation of the phenomenon see: *Entanglement Explained* on page 185)

---

26 To be absolutely fair, the experiment which Bell described would allow 25% of the monkeys to choose the same fruit but, take it from me, this does not materially alter the paradoxical nature of the experiment.

Heisenberg's Uncertainty Principle

What is the status of Heisenberg's Uncertainty Principle? Is it a consequence of the ambivalent nature of reality or is it an independent principle? How is it related to the theory of suspended reality and chaotic collapse?

Let's consider a case where the uncertainty principle can be used to calculate genuine results. Suppose a beam of electrons or a beam of light is directed at a narrow slit of width w. As expected, the beam will be diffracted and spread out in a fan on the other side whose breadth is typified by a semi-angle α. Standard wave theory predicts that
$\alpha \approx \lambda/w$ where λ is the wavelength of the electron or light.

It is possible to derive the same formula using Heisenberg's Uncertainty Principle. When the electron or photon whose momentum is P passes through the slit, its sideways position is confined to the width w. The Uncertainty Principle predicts that its sideways momentum p must therefore be uncertain to the extent of h/w. The typical angle of deflection is therefore approximately

$$\alpha = \frac{p}{P} \approx \frac{h/w}{P}$$

Now using the de Broglie relation $P = h/\lambda$

$$\alpha \approx \frac{h/w}{h/\lambda} = \frac{\lambda}{w}$$

This 'proof' has a number of deficiencies. First, it only gives an approximate answer and gives no hint as to how the expected intensity varies across the spread of the beam. Second, it seems to me to put the cart before the horse. One must not think that, just because the position of the particle is suddenly constrained to a width w that it is given a random sideways kick of magnitude p. Rather one should say that, having limited the range of available wavelengths associated with the sideways motion of the particle to a maximum of w, the sideways momentum of the particle must be of the order of p. The narrower the slit, the smaller the range of allowed wavelengths and the greater the range of allowed momenta.

We see, therefore, that Heisenberg's Uncertainty Principle is not

169

really a fundamental principle at all, but a *consequence* of the essentially wave-like nature of particles such as electrons and photons. It does not imply that there is any inherent randomness in QT. For a discussion of how the principle can be explained in terms of wave packets see ***The Wave-packet Theory of Light*** on page 162)

On the other hand, I believe that there *is* some inherent randomness in QT. What is more, the theory of chaotic collapse points out exactly how and where this randomness arises.

When a single electron passes through a classic double slit arrangement, the standard methods of quantum mechanics predict with great accuracy the *probability* that it will land in a particular region of the screen but where it *actually* lands is, apparently, completely random. This is because, when the electron (or its many manifestations while it is in suspended reality) begins to interact with the atoms and molecules in the screen, the wave function becomes unstable and collapses into a single reality. Which possibility is actually realized is determined by the initial conditions and by minuscule variations in these conditions due to the graininess of space time on the scale of the Planck length. (For more detail on this see: ***Chaotic Collapse*** on page 177.)

The Elitzer/Vaidman Bomb-testing Experiment

In the course of rationalising the Russian military after the collapse of the USSR, the authorities come across a secret store of bombs hidden away underground in a heavily fortified bunker. The documentation says that they are extremely sensitive and a single photon reflecting off the mirror attached to each bomb will recoil from the impact sufficiently to set it off. In order to test one of the bombs, they try out the experiment. In a dark underground chamber they direct a single photon at the mirror on the bomb and, sure enough, it explodes.

Just to make sure, they try another bomb – but this one fails to explode. Further examination reveals that it is not the bomb which is at fault; it is the trigger mechanism which, over the years has got rusty and the mirror is stuck fast so that it cannot move when the photon hits it.

Obviously it is no good testing each bomb in turn to find the good ones so a brilliant scientist comes up with the following idea using what is called an interferometer. Here is a diagram of the arrangement:

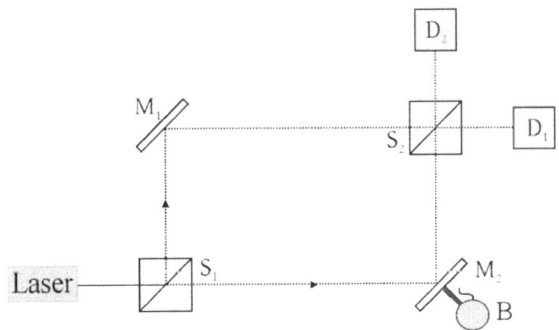

B is the bomb. M_1 and M_2 are mirrors (the latter being attached to the bomb). S_1 and S_2 are half-silvered mirrors. S_1 splits the beam of light from the laser (which in this case will consist of a single photon) into two and S_2 recombines the beams back into one again. D_1 and D_2 are photon detectors.

The interferometer is set up very carefully so that, under normal circumstances (i.e. when the mirror M_2 is rigidly fixed) light from the laser travels exactly equal distances in going from S_1 to S_2 by either route and in consequence it emerges from the second half-silvered mirror in the direction of **D₁**. The other detector **D₂** remains silent.

Let me repeat what we have agreed:

**** If the bomb is a dud, **D₂** will *never* fire. ****

Now if, as in the case of a live bomb, the mirror **M₂** is capable of wobbling, the interference between the two beams which occurs when they recombine is destroyed and half of the light will emerge towards **D₁** and half towards **D₂**. Naturally, the wobbling of the mirror will cause the bomb to explode.

But what happens if we send just a *single* photon through the apparatus? Quantum theory (as well as common sense) tells us that if the mirror on the bomb is fixed (i.e. the bomb is a dud), interference will occur and detector **D₂** will be silent. We are not allowed to ask which arm of the interferometer the photon went through because its *wave function* takes both paths.

On the other hand, we could, if we so wished *detect* which of the two paths the photon takes (e.g. by placing another detector – call it **D₃** – in one of the paths) The problem is that this obviously destroys the interference effect at **S₂** and detectors **D₁** and **D₂** will fire with equal frequency. In short, **D₃** will fire 50% of the time, and detectors **D₁** and **D₂** will fire 25% of the time.

Here is a picture of the arrangement I have in mind.

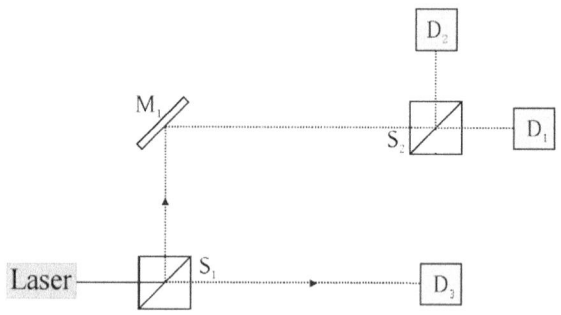

Let me repeat what we have just agreed. If **D₃** does not fire, then, sometimes at least **D₂** will.

172

Hang on a minute! What is our bomb if it isn't precisely that – a device which detects the presence of a photon?

So let me translate the last sentence into bomb language:

**** If the bomb is live, D_2 will *sometimes* fire ****

Now look at the two sentences which I have highlighted with asterisks. It is an logical deduction that:

**** If D_2 fires, the bomb is live ****

It is also true that, since we have only sent one photon through the apparatus, if D_2 fires, it cannot have been detected by the bomb i.e. the bomb does not explode.

The upshot of all this is that if we test 100 live bombs, 50 will explode, 25 will look like duds (but we can test these again later) and 25 will not explode *but we will know that they are live*!

We seem to have pulled of an impossible trick here. We seem to have gained information about an object without disturbing it in any way at all. We have just used the *potential* ability of the bomb to disturb the photon on its way through the interferometer without using its *actual* ability.

Well, believe it or not, this experiment has actually been carried out (though not with actual bombs of course!) and the prediction has been confirmed. What a strange world we inhabit!

The Logistic Equation

The simplest example of chaotic behaviour lies in the behaviour of the following innocuous pair of equations:

$$x' = A x y$$
$$y' = 1 - x$$

where $x + y$ always equals 1. (We can simplify this equation to just one variable by eliminating y getting $x' = A x (1 - x)$ but I have a good reason for writing it in two halves.)

Suppose that $A = 2$. Now think of a number between 0 and 1 – try ¼ for example. Put $x = ¼$ and $y = ¾$ into the formula and see what you get. Did you get x' 3/8 and y' = 5/8?

Now put $x = ⅜$ and see what you get. Keep putting the number you calculate back into the formula. What happens? The number gets closer and closer to the value $x' = 0.5$.

Lets see why this is so. If you put $x = 0.5$ into the equation you will find that (with $A = 2$) $x' = 0.5$ as well. Could we have predicted this? Yes. We obviously need to find a number x_0 such that

$$x_0 = A x_0 (1 - x_0)$$

We can easily rearrange this equation so get (either $x_0 = 0$ or)

$$x_0 = 1 - \frac{1}{A}$$

Try a few more values of A between 1 and 3 to see if the formula works. Now try $A = 3.3$. (Start with an initial number x of about 0.5 to save a bit of time) This should converge on the number 0.697. But it doesn't, does it? It oscillates between the values 0.479 and 0.824. The reason is that, while the number 0.697 is indeed a possible solution – it is not *stable*. Any slight deviation from this number diverges rapidly to a stable oscillation between the other two numbers.

To work out whether a particular solution x_0 is stable or not, we have to see if the *difference* between x' and x_0 increases or decreases with each iteration. If x_0 is a stable solution, $|x' - x_0| < |x - x_0|$ (where the bars indicate 'the modulus of' or, if you like, 'the absolute value of')

Now let us write

174

$$x = x_0 + e \quad \text{and} \quad x' = A(x_0 + e)(1 - (x_0 + e))$$

from which we obtain

$$|(A(x_0 + e)(1 - (x_0 + e))) - x_0)| < |e|$$

Putting $x_0 = 1 - 1/A$ we get after some rather messy algebra

$$|2 - A| < 1$$

This gives us what we require. The function is stable provided that A lies between 1 and 3. Between 3 and about 3.45, the function splits into two and beyond 3.45 it bifurcates again and again until eventually it becomes completely chaotic (though not without strange islands of stability here and there). Beyond 4, the function diverges without limit. This behaviour of the logistic equation is mirrored by a wide variety of physical systems which are stable up to a point, then start oscillating uncontrollably.

If you plot a graph of the stable points against A, this is what you get.

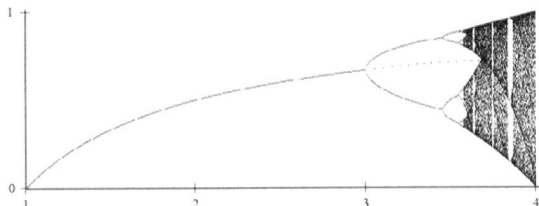

The aspect of this behaviour which has relevance to the theory of suspended reality is what is known as its 'sensitivity to initial conditions'. When A lies between 1 and 3, it doesn't matter what number you start with, the iterations will always home in on $1 - 1/A$ but if A is close to 4, the iterations diverge exponentially. Here is a simple spreadsheet showing this effect.

175

n			n			n		
0	0.100000	0.100001						
1	0.351000	0.351003	11	0.115748	0.115243	21	0.899929	0.596061
2	0.888416	0.888420	12	0.399167	0.397653	22	0.351220	0.939012
3	0.386618	0.386607	13	0.935348	0.934148	23	0.888672	0.223348
4	0.924864	0.924854	14	0.235842	0.239910	24	0.385843	0.676508
5	0.271013	0.271045	15	0.702861	0.711178	25	0.924176	0.853495
6	0.770504	0.770561	16	0.814505	0.801075	26	0.273292	0.487660
7	0.689628	0.689507	17	0.589237	0.621480	27	0.774553	0.974406
8	0.834760	0.834940	18	0.943943	0.917446	28	0.681020	0.097261
9	0.537949	0.537480	19	0.206367	0.295381	29	0.847204	0.342427
10	0.969384	0.969521	20	0.638740	0.811711	30	0.504853	0.878165

A is 3.9. We start with two numbers which differ by a fraction of 1%. After just 10 iterations, the discrepancy between the numbers has moved up 2 places of decimals and by the time we reach about 22 iterations, all correlation between them has been lost.

The rate at which the discrepancy increases is exponential and is characterised by a quantity called the Lyapunov exponent. If you concentrate on the last couple of places of decimals you will see that the discrepancy between them approximately doubles with each iteration. (This implies a Lyapunov exponent of about 0.7 because $e^{0.7} \approx 2$).

Now let us suppose that a discrepancy exists in a physical system which amounts to no more than the Planck length of 4×10^{-35} m (See: *The Planck Units* on page 178) If the discrepancy were to double every iteration, how many iterations would be needed to make the discrepancy one whole metre? The answer is only about 114 [27]. You see, once an exponential increase gets going, there is absolutely no stopping it.

---

27 $2^{114} = 2 \times 10^{34}$. This number of Planck lengths is about 1m.

Chaotic Collapse

Here is a very simple mathematical process involving three variables x, y and z which illustrates some of the features required of a mathematical theory of chaotic collapse. Just as we did with the logistic equation, we take the three variables and repeatedly plug them into the formulae to get new values and, as before, the equations are chosen so that the sum of the three variables is always equal to 1. Here they are:

$$x' = x(x+2y)$$
$$y' = y(y+2z)$$
$$z' = z(z+2x)$$

To save you a lot of effort, here is a graph showing how the three variables change over about 600 iterations:

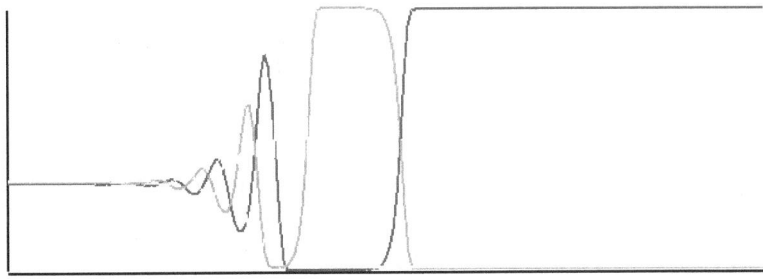

Initially, the three variables are very nearly (but not exactly) equal. Soon some wiggles develop and it is clear that the situation is unstable. Eventually, one or other of the variables hits either the ceiling or the floor. It is easy to prove that none of the variables can ever reach either exactly zero or exactly 1 (To do this requires that one of the other variables is already either less than zero or greater than 1.) but they can come arbitrarily close. If, however, the quantities x, y and z are granular (i.e. can on;ly take on integer multiples of a minimum quantity) then at some point two of them will become exactly zero and there will be only one winner.

The Planck Units

Planck's constant h has the units of *energy × time*.

Newton's Gravitation constant G has the units of *energy × length / mass²*

The speed of light c has dimensions *length / time*.

You can easily see from this that the quantity hc/G will have the dimensions of *mass²* or that $\sqrt{hc/G}$ will have the dimensions of *mass*. Given the accepted value of the three constants, this works out to be approximately 5.4×10^{-8} kg or the mass of a speck of dust.

Since *mass* has the same dimensions as *energy / velocity²* we can say that G has the dimensions of *length × velocity⁴ / energy*. This means that the quantity Gh / c^4 has the dimensions of *length × time* and the value 5.4×10^{-78} m s.

If we multiply this by c and take the square root, we will get a quantity with the dimensions of *length* and this works out to be 4×10^{-35} m. This is the Planck length.

If we divide it by c and take the square root, we get a quantity with the dimensions of *time* which works out to be 1.3×10^{-43} s. This is the Planck time.

(Note, if you use the *reduced* Planck constant $\hbar = h/2\pi$ you get slightly different values.)

The Planck units are considered to be fundamental in the sense that they are determined solely by the three fundamental physical constants which together describe Quantum Theory (h), Gravity (G) and Relativity (c).

It is often supposed that the Planck units of length and time determine the ultimate graininess of space-time and are the smallest quantities of length and time which have any meaning.

Obviously the same is not true of mass though because the Planck mass is quite large on an atomic scale.

Broken Symmetry

If a physical system is described by a linear differential equation, its future behaviour is unique (See: ***Determinism in differential equations*** on page 180). But not all physical systems behave like this. Consider a slightly flexible rod placed between the jaws of a vice (parallel to the axis of the vice). As the jaws are gradually closed, a simple differential analysis of the situation will suggest that, as the stress in the rod increases, so will its elastic strain without limit. This is not, in fact, what happens. At a certain point the lateral condition of the rod becomes unstable and it suddenly fails by bending in one direction or another. What determines the direction in which the rod bends? The answer is minute inhomogeneities in the crystal structure of the metal.

Another example is the beautiful crown of drops formed when a spherical object is dropped into a perfectly stationary fluid.

Since the initial conditions are perfectly symmetric, it is difficult to see why drops are formed in some places and not others. Here it is the force of surface tension which breaks the infinite symmetries which exist at the start down to 12 or so.

Other examples include the formation of hexagonal columns in shrinking basalt as at the Giant's Causeway and even the role of gravity in the formation of stars and galaxies. Symmetry breaking in the real world is obviously of crucial importance to our understanding but the subject has received little attention until now because it is mathematically so intractable.

Determinism in Differential Equations

One of the most important theorems in calculus states that for a broad class of linear ordinary and partial differential equations, given sufficient boundary conditions. there exists one and only one unique solution. For example. given the differential equation:

$$m\frac{d^2 y}{dt^2} - \sigma\frac{dy}{dt} + k\,y = 0$$

and the information that dy/dt and y have certain values at t = 0, the subsequent behaviour of y is completely determined for all t. (The equation is that which describes damped simple harmonic motion.) It is also true to say that the same equation and boundary conditions determine the complete history of the system as well – assuming, of course, that the differential equation has always applied to the system in question in the past. Now differential equations have been incredibly useful in forming a mathematical description of nature and this is the case for precisely this reason: We assume that the world we inhabit has a unique past and a unique future and linear differential equations have a unique past and a unique future too. But there are two reasons why we should doubt the assumption that the future is unique.

In the first place, the rigorous determinism inherent in classical physics as described by the differential equation is incompatible with our conviction that we have the power to alter the future if we so wish. Determinism is incompatible with Free Will.

Second, quantum theory seems to be telling us that there is an inherent randomness built in to the fundamental structure of the universe and there is no room for randomness either in a differential equation.

Now Schrödinger's wave equation is a perfectly standard linear partial differential equation and is therefore subject to the fundamental theorem. It follows that Schrödinger's wave equation cannot be a complete description of the universe. It must, of necessity, break down at the time when the wave function 'collapses' and if we are to make any progress in explaining the way the world works, we need to look first at ways in which Schrödinger's equation might be modified or extended to explain this process. (See: *Chaotic Collapse* on page 177)

The Quantum Zeno Effect

There is a well known (but little understood) quantum mystery which may (or may not) have a bearing on the issue of whether wave function collapse really happens or not.

The basic idea is this. Suppose you have a radioactive atom which you expect to decay with a half life of about 10 s. Periodically, e.g. every second, you give it a burst of photons from a laser and use the reflected photons to measure the mass of the atom and hence determine whether or not it has decayed. According to orthodox quantum theory, the 'measurement' will cause the wave function to collapse and the atom's internal 'clock' will, as it were, be reset to zero resulting in the atom being less likely to decay.

Experimental verification of this effect has been claimed but it is by no means universally accepted as being established. Nor is it clear whether the effect, if confirmed, would support or refute the various orthodox interpretations of quantum theory. My feeling is that ardent supporters of any interpretation would be able to interpret the data in whichever way they desired.

According to the suspended reality theory, the reflected photons would indeed collapse the wavefunction every second but I do not see why this would necessarily alter the atom's rate of decay. After a period of 1 second, the atom finds itself in a state in which it is entangled with an alpha particle. The wavefunction that describes this state can be divided into two with a probability of about 93% undecayed and 7% decayed components[28]. The burst of light collapses the wave function producing the expected rate of decay.

---

28 If the probability of decay in 1 second is p, the half life t is given by the equation $p^t = 0.5$. This means that $p = \sqrt[t]{0.5}$ or in this case with $t = 10$, $p = 0.93$. Hence the figures 93% and 7%.

The Wave/Particle Duality Explained

According to the Suspended Reality interpretation, entities like electrons and photons are real objects with real properties at all times. However, when a system is in a superposition of states (i.e. when reality id 'suspended') the values of these quantities may be different in different trajectories. The probability of each of these different trajectories is determined by a wave function similar, but not identical to, Schrödinger's equation. When the entity interacts with a macroscopic object, all but one of these trajectories leaving just one reality behind.

So while it is true to say, in retrospect, that the photon which was detected went through this slit, not that one, it is also true to say that, because the trajectories of other photons went through the other slit and contributed to the subsequent development of the wave function, *something wave-like* went through both slits and helped to determine where the photon was allowed to land. To put is crudely, you could say that during a period in which reality is suspended, the entity turns into a wave; but when chaotic collapse occurs, the entity turns back into a particle complete with all its past history. So after the event, the only reality which remains is that it was a particle all along!

In some respects, therefore, the Suspended Reality interpretation is similar to the Pilot wave interpretation. (See: ***Pilot waves and Virtual Particles*** on page 163.) but it does not require anything like 'spooky action at a distance' because the different trajectories evolve separately, only interfering with each other through the medium of a wave function.

Schrödinger's Cat Paradox Explained

The problem here is *superposition*. How can an electron be in two places at once? How can a cat be both alive and dead? According to the theory of suspended reality, these are two quite different questions.

In the first place, you will be glad to hear that a cat cannot be both alive and dead. The theory of suspended reality includes a mechanism which almost certainly guarantees the collapse of the wave function whenever more than a handful of atoms become entangled together.

The answer to the first question is rather different. An electron can *never* be in two places at once because it is a particle and particles always have a definite position etc. When we say that an electron, having passed through a double slit, is in a superposition of states, we are merely saying that the electron could be in one of two places but Nature has not yet decided which. When the wave function eventually collapses, the electron is revealed as having actually passed through one of the slits and not the other.

In 1999 an double-slit experiment was carried out using C_{60} molecules ('buckyballs') and in April 2011 it was announced that interference had been observed in a beam of molecules containing 430 atoms. A Bose-Einstein condensate is a large collection of atoms, cooled to a very low temperature which share the same quantum state. The largest such collection created to date contains more that 100 million atoms. Although not strictly an example of quantum superposition, it illustrates that it is possible for almost macroscopic objects to exists as a wave function without necessarily collapsing in an instant. There are many physicists of the Copenhagen or Many-Worlds persuasion who will claim that these observations show that there is no limit to the size of object which can, in principle, be placed into a superposition of states. I agree – *in principle*. But the difficulties seem to increase rapidly in practice, which is just what the theory of chaotic collapse would lead us to expect.

The Measurement Problem Explained

The theory of suspended reality and chaotic collapse solves the measurement problem instantly. There is no need to categorise some phenomena as 'measurements' and others not; there is no need to invoke 'consciousness' or even an 'observer'. Ultimately, the process of chaotic collapse of the wavefunction is no different from the propagation of the wave function; it is all part of the same process whereby the wave function develops according to well-defined (but at present unknown) and partially random rules.

It could be said, therefore, that under the Suspended Reality Interpretation, the only thing that actually exists at the present time is the wave function because, even as the wave function is undergoing chaotic collapse, it is also starting off on a new path with multiple possible trajectories leaving behind a history littered with particles like a rocket leaving behind a shower of sparks.

Entanglement explained

Although many see the measurement problem as being the central mystery of quantum theory, it was the issue of entanglement and 'spooky action at a distance' which suggested the idea of suspended reality to me. It seemed to me to be inconceivable that a photon on one side of the solar system could alter the behaviour of an entangled photon on the other side just because some property like its polarization had been measured.

Then when I read about the Elitzer/Vaidman bomb testing experiment it became obvious that even spooky action at a distance could not explain that. (How could the photon, having decided to fall on detector D_2 send a message back to the bomb to tell it *not* to explode? (See page 171))

This has been followed by a spate of new 'thought experiments' such as the Hardy-Jordan 'Dutch Door' experiment[29], all of which predict flagrant violations of common sense. All of them are, however, easily and naturally explained by the theory of suspended reality.

For example, when a pair of entangled photons whose directions of polarization are known to be parallel (e.g. see the experiment described on page 157) are sent to the edges of the solar system, the wave function which describes them contains information about all possible states of polarization[30]. When one of them is measured by passing through a polarizer at an angle θ the wave function collapses into the only state which remains possible – the state in which both photons had the same polarization as the polarizer all the time. There is no need for a message to be sent across the solar system – the trajectories which describe the other possible outcomes simply disappear. This is no more surprising than the fact that, when the winning number is drawn out of the raffle box, all the tickets that you hold in your hand suddenly become worthless. There is no need for

---

29 See: **Does Nature Violate Local Realism?** David Branning. *American Scientist. Volume 85. Sigma XI-The Scientific Research Society: Mar/Apr 1997* www.omnilogos.com/2011/12/13/does-nature-violate-local-realism/

30 In this respect, the theory of suspended reality differs significantly from the standard Copenhagen interpretation. In the latter, the entangled photons are regarded as being in one state whereas the former considers the photons to be in an infinite number of possible states simultaneously.

someone to go round the room crossing out all the losing tickets, they instantly become just waste paper.

When an electron, after passing through a double slit, hits a fluorescent screen, its wave function collapses to a state which is not only consistent with, but actually necessitates it having passed through one slit, not the other.

Similarly, the photon in the bomb testing experiment, after it has been through the first beam splitter enters a state of suspended reality. One of the halves of the wave function encounters the mirror attached to the bomb. If the bomb is a good one, this encounter causes the wavefunction to collapse and the bomb (potentially) to explode. But remember – this is only one of the halves of the photon's reality so the function can collapse in one of two equally probable ways. Either the the photon *really* explodes the bomb or it doesn't. If the latter, the suspended photon becomes real and finds itself travelling towards the second beam splitter where it divides into two again in the usual way.

If the bomb is a dud, the two suspended photons travel through the apparatus, interfere at the second beam splitter and hit detector D_1. In this case we do not know which way the photon went and we cannot be sure whether the bomb is live or not. We can, however, be sure that the real photon went one way or the other, not both.

If this is just too confusing, consider *The Twin monkey Paradox* described on page 167. You may remember that the monkeys choose what fruits they like in such a bizarre way that there is no explanation consistent with the hypothesis that they carry fixed genes from birth and it seems as if the decision made by the first monkey as to whether or not to accept the proffered fruit affects the decision made by the second monkey.

According to the theory of suspended reality, when the twins are born, the genes they carry are *undecided*. If you like, the universe splits into 8 copies in which the twins carry each of the 8 possibilities AAA, AAR, ARA, ARR, RAA, RAR, RRA and RRR (where each gene determines the respective propensity to choose a banana, an orange or an apple in that order). When the first monkey makes his choice (say he accepts a banana) all the trajectories disappear except one, namely the one in which both monkeys had the gene ARR. This ensures that his twin will inevitably accept a banana but reject either an orange or an apple. If the first monkey decides to reject the banana, all the trajectories disappear except the one in which the monkeys carry the gene RAA.

At no point do the monkeys have to communicate with each other, nor do they have to see into the future. The only counter-intuitive hypothesis that we have to accept is that the universe can, for short time at any rate, be in a state in which several different possible trajectories co-exist and that reality only exists in retrospect.

Coherence Length explained

According to the 'wave packet theory of light (see page 139), a photon has a definite length and contains a definite number of waves; something like this in fact:

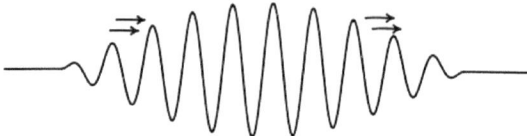

We can even attempt to measure the length of a photon using a simple interferometer as in the following diagram:

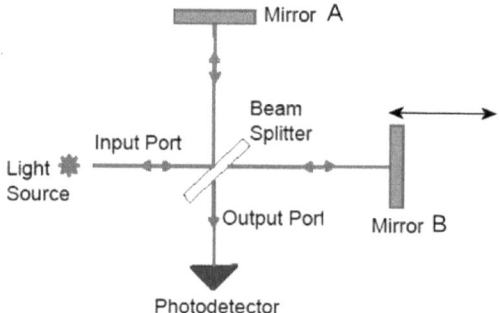

A beam of individual single photons is split into two potential beams which travel different distances to the two mirrors A and B. These mirrors reflect the potential photons back to the beam splitter which recombines them into a single beam again. If the difference between the distances which the two potential photons travel is a whole number of wavelengths, the photodetector will detect a photon, but if the distances differ by half a wavelength then destructive interference will occur and the photodetector will see nothing. (For more information about interference see: *The Double-slit Experiment* on page 146.)

Now as mirror B is moved further and further to the right, the photodetector sees light – dark – light – dark etc. as the two waves shift in and out of phase. This works fine up to a point, but eventually, when the difference in the path lengths is about a metre, the interference pattern fades and the photodetector simply responds to half the photons released. The *coherence length* of a photon is said to be about 1 m.

On the face of it, this seems to be powerful evidence that wave packets really do exist and that they are about a metre long, but this conclusion is not justified. Although the idea of coherence length is useful, it does not really tell us anything about the true nature of photons; it tells us more about the nature of the apparatus we use to observe them. According to the theory of 'Suspended Reality', coherence length is explained as follows:

When photon is emitted from an atom it enters a period of suspended reality in which its coherence length is *undefined*. You can, if you like, think of there being an infinite number of possible wave packets, some short some long all of which exist at the same time.

Now if the path difference in the interferometer is only a few wavelengths, it doesn't matter how long the coherence length is – they will all exhibit the same interference pattern. But if the path difference is, say 1m, then those potential wave packets that are shorter than 1m will not exhibit interference because they can only interact with each other in regions where they overlap. We might, however, expect to see a certain amount of interference from those potential wave packets that are longer than this. In practice, however, the interference pattern has completely disappeared by this time. The question then arises – why are there no potential wave packets longer than about 1 m? Is there a physical limit to the length of a wave packet? If so, what are the laws which determine this limit?

The answer is that there is no *physical* limit to the length of a potential wave packet; the limit is imposed by the dimensions of our apparatus, in particular, the distance from mirror A to the beam splitter. If mirror A is of the order of a metre or so away from the beam splitter, then the wave packets which can cause interference cannot be longer than this distance and the interference will disappear when the path difference becomes greater than this.

Emergent and Transcendent Properties

Science owes much of its success to its fundamentally reductionist approach which takes as axiomatic that if you understand the detailed workings of all the parts of a system, you understand the whole. Conversely, if you want to understand the whole system, you must understand the workings of all its parts.

In practice, however, this happy state of affairs is very difficult to achieve and much understanding of complex systems has been gained by introducing concepts which, only later perhaps, become understood at a deeper level.

When Robert Boyle and others discovered the laws which govern the behaviour of gases, he had no idea why gases behaved in this way and it wasn't until Daniel Bernoulli laid the foundations of the kinetic theory of gases that the laws could finally be understood at a fundamental level. We can now see clearly that the ideal gas equation PV/T = constant is a prime example of what is called an *emergent phenomenon* and that pressure and temperature are *emergent properties* of a gas. We can also see that it is perfectly possible to carry out important research into emergent phenomena without knowing how these phenomena and properties relate to lower levels of understanding. All you have to assume is that either someone else understands the relationship or, failing that, the relationship exists and someone will work it out in the future.

For example, a chemist can carry out research into the way certain molecules react without himself understanding the laws of physics which govern the interaction between atoms and electrons; likewise the biologist can carry out research in genetics without necessarily understanding the chemistry of DNA; a meteorologist can predict the course of a hurricane without using Newton's Laws; a psychologist can study animal behaviour without necessarily understanding how neurons work etc. etc. Indeed, without emergent properties like hurricanes and genes, 99% of science would be totally impractical.

Some systems are so complex that they can be understood at several different levels simultaneously. The humble PC on your desk is a prime example of what I mean. Suppose you are asked 'How does it work?' I can think of at least six different answers.

The computer games freak might reply: "Well, if you pick up a Zeta-blaster with the mouse and fire it at the Zombie, you can blast the hell out of him like this."

The geek who wrote the program might reply: "All the objects that appear to you on the screen are in fact instantiations of certain classes which have various numerical properties like position, velocity etc., and methods which describe the way they react when they encounter other objects."

The computer specialist who wrote the operating system will say: "The program you are running is just a vast array of binary numbers stored in the computer's memory. These numbers are being constantly updated by the CPU which is just carrying out a series of logical operations on them at a rate of about 2 billion operations every second."

The engineer who designed the chip will, however, say that, fundamentally the way it works is that electric potentials and currents are causing changes in the state of the transistors in the logic gates which comprise the CPU and the computer's memory chips.

A physicist would point out that electric potentials and currents that flow through the transistors can only be fully described in terms of the interactions between the electrons and the atoms in the devices using quantum theory.

In truth, of course, they are *all* right. But each explanation is only valid on the level at which it applies and the PC is a marvellous example of a system which can be explained at many different levels, each with its own set of emergent properties and phenomena related to and understood in terms of properties and phenomena at a lower level.

But is this always the case?

At any stage in the development of a branch of science there are times when phenomena are discovered which cannot be explained in terms of concepts at a lower level and a good example is the Second Law of Thermodynamics. (See: ***The Second Law of Thermodynamics and the Arrow of Time*** on page 114.) Here we have a simple straightforward law which cannot be deduced form Newton's Laws of Motion for one very simple reason. The Second Law is asymmetric with respect to time but Newton's Laws are not. As long as you believe in the fundamental correctness of Newton's Laws, you cannot explain the Second Law. The Second Law *transcends* Newton's Laws and it cannot be described as simply an *emergent phenomenon* – it is a *transcendent phenomenon*.

(If, however, you believe as I do that interactions on a minute scale are subject to randomness at the quantum level, then the Second law becomes explicable and its status reverts to being emergent rather than transcendent. This idea is explained more fully in the section referred to above.)

We can now put forward a definition of these two concepts more clearly. A *emergent phenomenon* is a phenomenon which uses concepts and obeys laws which we are confident can be fully explained in terms of more fundamental known concepts, if not now, at least potentially in the future. A *transcendent phenomenon* is one which appears to be inexplicable in terms of more fundamental concepts as they are currently understood either because the fundamental laws appear to be totally inadequate to explain the phenomenon or because the phenomenon and the fundamental laws appear to be talking about completely different things. The case of Newton's Laws being totally inadequate to explain the Second Law of Thermodynamics is a good example of the former and the levels of description of the workings of a PC afford us an excellent example of the latter.

The patterns of pixels on the screen can be directly related to numbers held in memory which correspond exactly to voltages and currents in the wires which in turn obey the laws of physics. So we can, in principle at any rate, explain why this pixel on the screen is coloured blue by relating it all the way down to the behaviour of the electrons in the electrical circuits. But there is no way we can explain why 'Zombies' are 'killed' when you zap them with a 'Zeta-blaster'. The whole concept of a 'Zombie' or a 'Zeta-blaster' only makes sense in a context which includes things like the Vietnam War and films about alien space invaders. You can't explain all the properties of a Zombie in terms of electrical voltages. Zombies are cultural artefacts and lie outside the realm of physical explanation. It is no slur on the power of Physics to say that it cannot explain the behaviour of Zombies; it is simply that the behaviour of Zombies is dependent on factors other than physical ones.

Many other examples can be found in the social sciences (why are women paid less than men?) economics (why does increased supply depress prices) and the Arts (what are the defining characteristics of a piece Baroque music?). The concepts of women, prices and music simply have no meaning at lower levels and cannot therefore be explained in reductionist terms.

The big question now arises – which of the unsolved mysteries of science involve transcendent phenomena rather than mere emergent ones? Will the origin of life turn out to be a random chance coming together of millions of molecules to make the first strand of DNA or is will we need some wholly new principle to explain it? Will it really be possible to explain how a single fertilised ovum turns into a human being in chemical terms alone, or will we need some new ideas? Can the fact that human beings can remember and recall events from their childhood really be explained in terms of the way certain neurons are wired up together or are we missing something? And above all, is the conscious brain to be explained solely in terms of neuronal activity or is consciousness beyond the reach of any reductionist explanation?

In my opinion the origin of life, morphogenesis and memory will probably turn out to be emergent and will eventually succumb to a reductionist explanation but I do not believe that a reductionist explanation of consciousness can ever be achieved because, like Zombies in a computer, consciousness *has no meaning* at the neuronal level. Even if we were able at some future time to look into a brain and by examining the activity of its neurons deduce whether or not it was conscious, we still would not have *explained* what consciousness is. You might be able, in a similar way, to look inside a computer and say that an instance of the class held in memory at location #123456 is currently displaying on the screen – but that will never *explain* what a Zombie is and why you need a Zeta-blaster to kill it.

A line-following robot

The diagram below is a circuit diagram for a line-following robot. The chip in the centre is nothing more than a dual amplifier which drives the two motors, each of which drive a single wheel on the side of the robot. The LDRs (light dependent resistors) are positioned just either side of the line and they are connected to the amplifier in such a way that when the LDR on each side detects light, the corresponding motor rotates forwards. If one of the LDRs strays over the (black) line, the motor on that side stops causing the robot to swerve back on course.

All that is needed to convert the robot to a white line follower is to swap the inputs on pins 2 and 7 on the left hand side and pins 15 and 10 on the right.

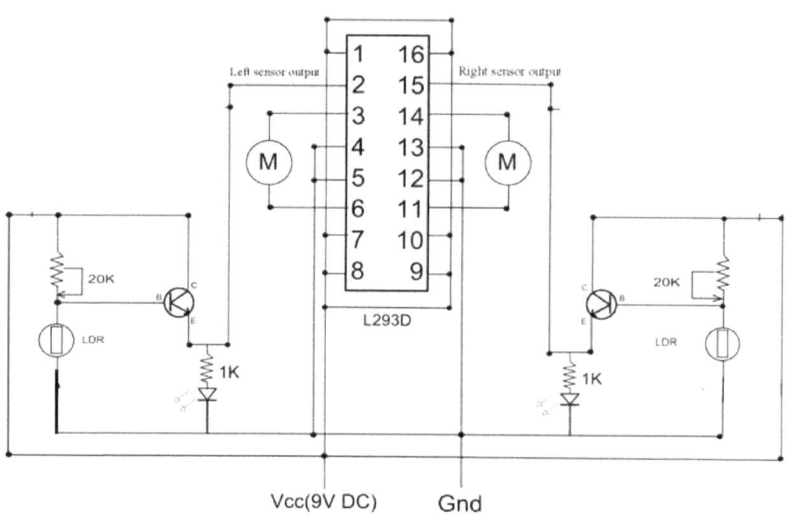

Adapted from a diagram published on www.societyofrobots.com

EEG patterns in sleep

A crude but important method of observing the activity inside the brain in both the conscious and unconscious states is with an electroencephalograph or EEG and up to 5 different characteristic wave patterns can be observed in the human brain while awake and asleep. (see figure below)

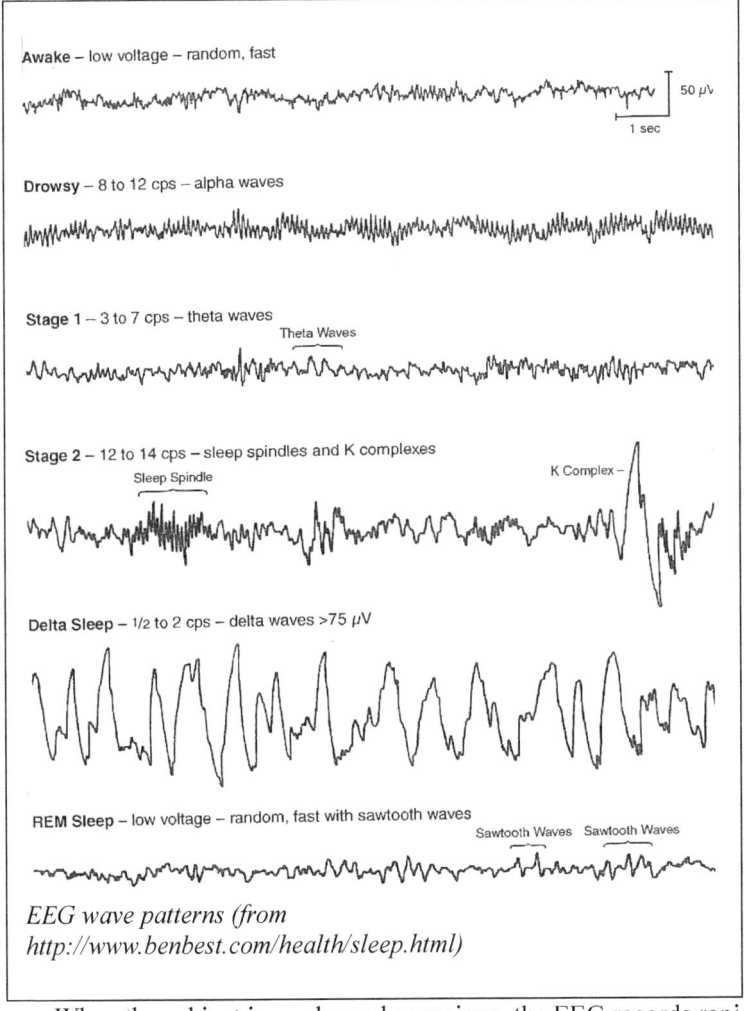

Awake – low voltage – random, fast

50 µV

1 sec

Drowsy – 8 to 12 cps – alpha waves

Stage 1 – 3 to 7 cps – theta waves

Theta Waves

Stage 2 – 12 to 14 cps – sleep spindles and K complexes

Sleep Spindle

K Complex –

Delta Sleep – 1/2 to 2 cps – delta waves >75 µV

REM Sleep – low voltage – random, fast with sawtooth waves

Sawtooth Waves   Sawtooth Waves

EEG wave patterns (from http://www.benbest.com/health/sleep.html)

When the subject is awake and conscious, the EEG records rapid and irregular oscillations of relatively small amplitude. Since the EEG electrodes cover a vast area of cortex (thousands if not millions of neurons) this observation is consistent with the hypothesis that, while the brain is awake, all areas of the brain are more or less active and are 'doing their own thing'.

As the subject begins to fall asleep, the famous 'alpha' rhythm starts to become evident. This is a more coordinated oscillation with a frequency of 8 to 12 Hz. This stage is known as 'drowsy sleep' and the subject may be conscious of his surroundings but unable, or unwilling, to react to them.

The first stage of true sleep is characterised by an even slower oscillation called 'theta waves' of frequency between 3 and 7 Hz. The subject is now truly unconscious (but see below) and over the next 20 minutes or so, descends into two further stages of deep sleep the last of which – delta sleep – is characterised by highly coordinated electrical oscillations of large amplitude and slow frequency.

At intervals during the night, the subject returns to stage 1 sleep and enters what is known as Rapid Eye Movement or REM sleep. The EEG pattern is similar to the waking state with rapid, low voltage oscillations and oxygen consumption by the brain increases dramatically but the subject's muscles are (usually) paralysed and he is more difficult to wake than at other times. As the name suggests, the stage is accompanied by rapid movements of the eye and if the subject is wakened during this phase, he is more likely to report that he was dreaming at that instant.

Penrose's Proof

During the first two decades of the 20th century, mathematicians, notably Alfred North Whitehead and Bertrand Russell, embarked on an ambitious program to show that Mathematics was a complete, logical watertight system in which every true statement could in principle be proved and every false statement disproved. But their hopes were dashed when in 1931 a young Austrian mathematician called Kurt Gödel published a theorem which turned Mathematics upside down. In effect what he discovered is that if Mathematics if consistent, it cannot be complete – that is to say, in any given mathematical system, there will always be true statements which cannot be proved *within the system*. What is more, these statements can be shown to be true by using logic *outside the system*.

Now it has to be said that, unexpected and disturbing as it was, this revelation has not stopped mathematicians going about their business, nor has it falsified any of the mathematics so far discovered; it has merely put a theoretical limit on what mathematicians can hope to achieve.

Sir Roger Penrose, however, has seen in Gödel's theorem a kind of proof that human brains must operate in a non-classical mode. He argues that any classical computer can, in principle, be programmed to prove all the provable theorems in any well-defined mathematical system. But by 'thinking outside the box' a human brain can also deduce the truth of some non-provable (Gödelian) statements as well. It follows therefore that the human brain must employ some non-classical processes.

It is not for me to defend or criticise this argument but I do agree with the conclusion, but for different reasons.

Critical Comments

Isn't suspended reality and chaotic collapse just another interpretation of quantum theory?

Suspended reality is but chaotic collapse is not. If it ever becomes possible to formulate the mathematics which leads to the collapse of the wave function, experiments could be devised to test whether, or not, it does occur. I suspect that hopeful designers of quantum computers will find it more difficult than they think to preserve their quantum states for sufficient lengths of time to do anything useful with them. If so it will become imperative to study quantum collapse experimentally. It is chaotic collapse which explains the element of randomness which is crucial to explaining the Second Law of Thermodynamics and which enables us, at least in retrospect, to give a sensible account of the past history of our world.

But leaving aside chaotic collapse, isn't the theory just another version of the standard Copenhagen theory dressed up in a bit of fancy language?

Yes – to a certain extent it is. The concept of suspended reality and different alternative trajectories is really just another way of saying that, during the period of suspended reality, many possibilities exist at the same time, just as a photon can be in a superposition of states when it has passed through a half-silvered mirror.

There is one important difference, however. In the standard interpretation, a quantum object such as a photon *does not have* a direction of polarization until it is measured. In the Suspended Reality interpretation, a photon whose polarization has yet to be determined is in a superposition of an infinite number of states, each with a different polarization. And because the wave function which describes this situation has to keep account of all these possibilities, it cannot be identical to Schrödinger's equation. In fact, I suspect that some new mathematics will be necessary to formulate it.

In the absence of a mathematical description of the precise mechanism of chaotic collapse, aren't you really saying that wavefunction collapse occurs when a measurement is made but you don't know why?

Yes, that's true too – but lots of phenomena happen without us being able to describe the detailed mathematics. Could we have predicted the rings of Saturn or calculate the caustic light curves on the bottom of a swimming pool in the sunshine? Some things are just too complicated. I am, however, offering the hope that some day we might be able to understand the process and, perhaps, to control it.

During the process of chaotic collapse you claim that the random increase in probability of the wave function at one point causes the instantaneous collapse of the function at all other points. Isn't this just 'spooky action at a distance' in disguise?

No. Absolutely not. The theory no more requires instantaneous communication than the deduction that, having found an odd sock in the washing machine, there must exist an odd sock in the drawer upstairs which is the same colour as the one you have just found.

It is true that one of the conditions of the theory is that the total probability of all the possible trajectories must be 1. But this is just a tautology. To misquote Holmes – when you have discovered the truth, every other possibility, however plausible, must be false.

Suppose you enter a raffle with 100 tickets for sale at £1 each with a prize of £100. Before the draw, the raffle tickets are worth £1 each and could even be used as currency. But as soon as the winning number is revealed, the ticket bearing that number is worth £100 and the others are instantly rendered worthless. There is no 'spooky action at a distance' here.

If, as you say, while reality is 'suspended' reality splits into multiple almost identical copies, isn't this just the Many Worlds interpretation with added collapse?.

Emphatically not. Consider what happens when a photon approaches a simple polarizer. In the Many Worlds interpretation (and the Copenhagen interpretation also) before the photon reaches the polarizer there is only *one* world in which the state of polarization of the photon is a meaningless concept. In the Suspended Reality interpretation, there are *infinitely many* worlds containing every possible orientation of the photon.

After the photon has been either absorbed or transmitted, in the Many Worlds interpretation there are now two parallel worlds, one in which the photon was transmitted, another in which it was absorbed. In contrast, in the Suspended Reality interpretation there is now only one real world which contains either a transmitted or an absorbed photon. What is more, in this world it was the case that the photon always was oriented in the direction which caused that result before it reached the polarizer. Reality is preserved at all times, only retrospectively

So in a sense the Suspended Reality interpretation is exactly the opposite of the Many Worlds interpretation.

Haven't I read somewhere about a new interpretation of quantum theory involving what are called 'consistent histories'? Isn't that what your theory is all about – setting up consistent histories of an event and then selecting one of them to become real?

I have to confess that I do not understand any of the language which the proponents of this interpretation use. Sometimes they claim that their idea is no more than a new slant on the Copenhagen interpretation; but then it is also claimed that the consistent histories approach denies the reality of wavefunction collapse, which seems to me to be a contradiction. What I can say is that in the suspended reality theory, all the possible histories of an event have to be consistent – i.e. each has to be *possible* – and it may well turn out that the mathematics of the consistent histories approach will be of use in constructing the new models required to formalise the details of chaotic collapse.

It is a pity that the phrase 'consistent histories' has already been used as I would have used it myself if I had thought of it first!

You have stated that Schrödinger's wave equation is completely deterministic and you have described the collapse of the wave function using the concept of deterministic chaos. Doesn't this mean that the universe you describe is still as deterministic as Newton's and the 'free will' is therefore still an illusion?

What you say would be true if space and time were continuous. It is commonly believed, however, and I concur with this view, that space and time are discontinuous on the scale of the Planck units. As soon as more than a few dozen atoms start interacting, the wavefunction becomes extremely sensitive to its initial conditions and random changes on this scale rapidly magnify to the point where collapse into one actual state becomes inevitable. But exactly *which* state is essentially random, only constrained by the probability of that trajectory as determined by the state of the wave function just before collapse.

But if the future is determined by random events, how can conscious beings alter the future in predictable ways?

I don't know. Nobody does. All I can say is that I believe that there are processes going on in the brain of which we have no current understanding and that these processes (which may or may not be quantum in nature) both give us the experience of consciousness and the ability to make informed decisions which affect the future.

Moving on to your claims about human consciousness, you seem to be saying that we don't understand consciousness and we don't understand quantum theory, therefore the answer to the former puzzle must be the latter.

Yes, I admit, it does look rather like that – and yet I think there is a connection. Whenever we think about the way the human brain thinks and how it differs from the way a computer works, I am always struck by the holistic nature of the brain as opposed to the particular nature of the computer. Take the brilliant Countdown competitor who can figure out the Countdown conundrum (a 9 letter anagram) almost before I have read the letters. He seems to be able to take all the letters in at once and come up with the answer instantly. Whatever is going on in his brain, he certainly isn't checking every possible combination of letters against a stored dictionary. Likewise a chess master may check out a few possible moves in his head but he certainly doesn't check out *all* of them in any systematic way as a computer would; he takes the whole situation in at once, as it were.

Now one possible difference between the way a computer works and the way a brain works is that a computer can (generally) only process one bit of information at a time while it is probable that a brain employs parallel processing on a massive scale. I do not, however, think that this difference is sufficient to explain the huge gulf that exists between a computer and a brain. Experiments with neural networks have produced some interesting results – robots equipped with them have learned to run mazes and learn languages – but it is a trivial task to simulate a neural network on a conventional computer. In principle a computer with a single CPU can do anything a neural network is capable of.

Another relevant point to make here is that the conscious brain is essentially a single-threaded device. You cannot think two thoughts at the same time. When two friends in deep conversation have to cross a busy road, the conversation stops until the tricky manoeuvre is complete. So while many of the brain's substructures can handle many processes in parallel, when it comes to conscious thought, the whole of the higher brain is involved.

All this convinces me that some holistic process is going on in the conscious brain which is wholly absent from any computer which we can currently build. Quantum theory, as I have interpreted it seems to include a holistic element in that the outcome of any quantum process has to take into account everything which is going on round it and I find this at least highly suggestive.

You have attempted to explain how conscious brains can exercise their free will by claiming that it is the holistic processes going on in the whole brain which affect individual neurons in a non-classical manner. You have also claimed that all quantum processes such as the decay of a radioactive atom in principle involve contributions to the collapse of the wavefunction from every other atom in the universe. Would it not be fair to say that, in a sense therefore, the whole universe is a conscious computer which exercises its free will whenever it chooses to make a radioactive atom decay?

I have no objection to you using this kind of language if you wish. You can even call this supposedly conscious entity God if you like. I prefer to restrict the use of the word conscious to structures of such exquisite complexity that vast numbers of atoms separated by distances of several centimetres make significant contributions to the eventual outcome. In the case of a radioactive atom, I feel sure that no more than a dozen of its nearest neighbours have any significant effect.

Do you think there is a realistic possibility of us ever creating a truly conscious computer?

I have to say that I am extremely dubious about this. It is conceivable that, in the not too distant future, we will be able to construct quantum computers which can perform many tasks simultaneously but just because a computer uses quantum phenomena does not mean that it is conscious any more than a radioactive atom is conscious.

If we are to build a conscious computer, we will first have to build machines which are capable of maintaining quantum states on a macroscopic scale for significant periods of time. Research into quantum computers may show us the way forward here. Secondly, we will have to understand in detail how chaotic collapse comes about; thirdly, we will have to understand exactly how quantum processes are used in a brain – a task that will involve monitoring brain activity on a minute scale; and finally, we will have to use our knowledge to design and construct our thinking machine, probably using nano-technologies undreamed of today. Even if all these programs were to proceed successfully, I can't see any prospect of a result for many, many decades if not centuries.

And there is the final consideration. Why should we want to build such a machine? Every man and woman knows a much easier and much more satisfactory way!

You claim that consciousness depends on some unknown and possibly unknowable process going on in the brain which makes you a dualist of a sort – at least a mysterian, but on the other hand, you reject the notion of an independent 'self'. If you are prepared to invoke what we might call 'miraculous' processes to explain consciousness, why are you so averse to accepting the 'miraculous' nature of an independent 'self'?

It is not fair to describe my 'unknown processes' as 'miraculous'. There is no reason to suppose that they are any more miraculous than the 'miraculous' way in which the Earth attracts an apple through the medium of gravity. If, as I suspect, our human brains will never be able to understand how these process bring about the experience of consciousness, they will continue to appear mysterious but that doesn't mean that they are outside scientific scrutiny.

The problem with an independent 'self' is two fold. In the first place the concept is fraught with paradoxes and, more importantly, 'selves' as conceived by true dualists are definitely outside the scope of scientific enquiry.